Alexandra Kelly Nothstein

Selenium Transfer between Kaolinite or Goethite Surfaces, Nutrient Solution and *Oryza Sativa*

Karlsruher Mineralogische und Geochemische Hefte

Schriftenreihe des Instituts für Angewandte Geowissenschaften,
Karlsruher Institut für Technologie (KIT)

Band 41

Selenium Transfer between Kaolinite or Goethite Surfaces, Nutrient Solution and *Oryza Sativa*

by
Alexandra Kelly Nothstein

KIT Scientific Publishing

Dissertation, Karlsruher Institut für Technologie (KIT)
Fakultät für Bauingenieur-, Geo- und Umweltwissenschaften, 2015
Referenten: Prof. Dr. Thomas Neumann, Prof. Dr. Peter Nick

Impressum

KIT Scientific Publishing

Karlsruher Institut für Technologie (KIT)
KIT Scientific Publishing
Straße am Forum 2
D-76131 Karlsruhe

KIT Scientific Publishing is a registered trademark of Karlsruhe Institute of Technology. Reprint using the book cover is not allowed.

www.ksp.kit.edu

Print on Demand 2016

ISSN 1618-2677
ISBN 978-3-7315-0461-0
DOI: 10.5445/KSP/1000051067

Karlsruhe Institute of Technology

Selenium transfer between kaolinite or goethite surfaces, nutrient solution and *Oryza sativa*

accepted in fulfilment of the requirements for the degree of

Doctor of Natural Sciences (Dr.-rer. nat.)

to the

Department of Civil Engineering, Geo- and Environmental Sciences

Dissertation of

Dipl. Geoökol. Alexandra K. Nothstein

born in Hof, Saale

Day of oral examination: 01.07.2015

Reviewer: Prof. Dr. Thomas Neumann
Second reviewer: Prof. Dr. Peter Nick
Advisor: Dr. Elisabeth Eiche

Karlsruhe, 31.08.2015

Acknowledgement and Funding

This thesis would not have been possible without the support of many helping hands – and minds. Therefore, I would like to use this opportunity to express my gratitude towards all of the people who have made these studies possible for me with helpful comments, practical advice and moral support.

First and foremost, I would like to thank my **supervisors and advisers** who patiently helped with all major and minor troubles accompanying this work, **Prof. Dr. Thomas Neumann**, **Prof. Dr. Peter Nick** and **Dr. Elisabeth Eiche**. This includes the financial support from **Dr. Monika Stelling's Young Investigator Group (YIG)** on 'Stable isotopes (Se, O) as source and process indicator'.

I would also like to thank the members of my PhD Commission for their support: **Prof. Dr. Thomas Neumann**, **Prof. Dr. Peter Nick**, **PD Dr. Katja Emmerich**, **PD Dr. Stefan Norra**, **Prof. Dr. Dieter Burger** and **Prof. Dr. Josef Winter**.

Furthermore, I would like to thank all colleagues, lab assistants and technicians at the AGW for the wonderfully professional, friendly and positive working atmosphere. There was never a moment in which I doubted I could ask anyone about anything related to my work. I would like to acknowledge in particular:

- **Claudia Mößner** for infinite problem solving strategies concerning analytical difficulties, IC and ICP-MS measurements
- **Gesine Preuss** and **Ralf Wachter** for patient technical assistance with countless HG-FIAS-related challenges
- **Cornelia Haug**, now retired, for practical advice on all topics from chemical waste disposal and acid distillation to developing a new way to safely clean quartz tubes with hydrofluoric acid
- **Beate Oetzel** for her help with XRD sample preparations
- **Kristian Nikoloski** for preparation of smelting beads for me

Special thanks go to **Dr. Katja Emmerich** (CMM-IFG, KIT) and **Dr. Annett Steudel** (CMM-IFG, KIT) for their help on material characterisation of my clay minerals, **Dr. Peter Weidler** (CMM-IFG, KIT) for the BET measurement of my sorption materials, to **Volker Zibat** (LEM, KIT) for SEM images of sorption samples as well as **Dr. Michael Riemann** (CFN, KIT) and **Rita Brendel** (CFN, KIT) for their practical help on rice cultivation, sterile working conditions & other biology involved.

I would also like to thank **Prof. Dr. Lenny Winkel** and her lab group with **Dr. Gerrad Jones**, **Dr. Elke Suess**, **Caroline Stengel**, **Bas Vriens**, **Tim Blazina**, **Boris Drotz** and especially **Katja Luxem** for the wonderful opportunity I had to discuss my work and to test an experimental set-up for Se-volatilisation.

Further valuable discussions on methodology, analytical problems and moral support have also been imparted to me through fellow PhD students at AGW, whom I would also like to thank here: **Yuan Chen**, **Helena Banning**, **Nicolas Börsig**, **Arno Hartmann**, **Andreas Holbach**, **Sebastian Potsch**, **Harald Neidhardt**, **Alexander Diener** and **Peter Illner**.

For direct contribution to my work, I would also like to thank the following diploma and bachelor students, who directly or indirectly helped to improve many steps of my experiments, analytical methodology and through whom I was able to learn valuable teaching skills: **Sara Ziegler**, **Gabriele Konrad**, **Matthias von Brasch**, **Christian Engelke**, **Ruben Kosmala**, **Philipp Maier** and **Arne Tenspolde**.

Some of my experimental work included special handcrafted workmanship, which is why I would also like to express appreciation to the head of the workshop, **Bernd Grunvinck** (now retired) and his team.

Furthermore, I would like to thank my brother, **Mark Nothstein**, for helping with my mathematical grasp on the modelling of my data and countless LaTeX issues.

And, of course, I would also like to express my special thanks and gratitude toward everyone who has directly improved my writing of this dissertation by proof-reading my many drafts:

- **Dr. Elisabeth Eiche**, who greatly improved my precision on technical details, methodical aspects and viable discussion points
- **Arno Hartmann**, who had questions no one else had and patiently listened to all my rants at random hours of the day when things didn't work the way I wanted them to
- **Mary Desik**, my mom, who put up with my constant misuse of prepositions and commas and unravelled countless lengthy sentences full of German grammar, reducing my art of convoluted sentence writing to a painless reading experience
- **Mark Nothstein**, my brother, who pointed out that I shouldn't begin with describing everything that didn't work

And last, but certainly not least, I would like to thank my family and friends for their invaluable support and continuous encouragement.

This work was funded by:

- a scholarship by the Landesgraduiertenförderung Baden-Württemberg at the KIT (Karlsruhe Institute of Technology)
- a scholarship by the Graduate School for Climate and Environment (GRACE) at the KIT (Karlsruhe Institute of Technology)
- coverage for analytical costs by Dr. Monika Stelling's Young Investigator Group (YIG), KIT
- coverage for the conference attendance of the 3rd International Selenium Conference in Hefei, China by GRACE, KIT
- coverage for a trip to the Advanced Photon Source (APS) at Argonne National Laboratory, USA for XANES and XRF measurements by GRACE, KIT
- coverage for analytical costs (ICP-MS and HG-FIAS measurements) by GRACE, KIT
- coverage for a 3-month research stay at the Swiss Federal Institute of Aquatic Science and Technology (Eawag) in Zürich, Swizzerland by GRACE, KIT

Abstract

Management of Se resources is a vital topic. This metalliod trace element is essential for many organisms and has a small range between human dietary deficiency (< 55 µg/d) and chronic toxicity (> 400 µg/d). Moreover, Se is unevenly distributed throughout the world and its bioavailability is governed by many factors. Biofortification and Se toxicity risk assessment are key aspects of managing global Se resources. Therefore, detailed research is required regarding biomolecular and geochemical interactions between Se and the environment. Accordingly, the aim of this study was to describe and quantify Se transfer in the Critical Zone, i.e. at the interface between soil and plant.

For this purpose, the Critical Zone was compartmentalized into three Se reservoirs: "soil", "soil solution" and "plant". Experimentally, these were represented by the model minerals kaolinite and goethite, a nutrient solution and rice plants (*Oryza sativa*), respectively. Using identical experimental parameters, first the speciation-dependent Se transfer from solution to plant and subsequent Se partitioning within the plant was studied; then adsorption-desorption processes of selenate and selenite onto the model minerals kaolinite and goethite were investigated. After that, both experiments were combined. Quantification of Se transfer between solution and mineral substrate as well as Se-uptake into plants from the solution provided the basis for a mathematical process model.

To determine Se-uptake differences due to lack of nutrients, a series of closed plant-box systems was used; *Oryza sativa* caryopses were prepared with Se-spiked nutrient-free phytoagar or nutrient solution of 0 - 2500 µg/L Se as selenite or selenate. Furthermore, forcing plants to grow through a Se-free agar-filled tube before reaching the Se-spiked nutrient solution or phytoagar made it possible to observe the effects of delayed Se treatment. All plants were grown with 14 days of Se-uptake in a climate chamber. After harvest, plant bulk samples of shoot and root were freeze-dried and microwave-digested with nitric acid and hydrogen peroxide. Selenium content was measured with HG-FIAS. Agar samples were digested similarly and measured with HG-FIAS. Selenium and nutrient content of the nutrient solution was measured by ICP-MS and IC.

To characterize Se adsorption-desorption behavior, a series of 24-h sorption batch experiments was carried out with 0.5 g kaolinite or goethite in a KCl-solution resembling the nutrient solution in ionic strength. Solutions were spiked with 0 - 5000 µg/L Se as either selenite or selenate. To study competition effects, these experiments were repeated in the presence of 750 µmol/L nitrate, sulphate or phosphate. To quantify ion exchange potential, subsequent Se desorption experiments were conducted with 0.1 mol/L K_2HPO_4.

To study combined effects of Se sorption processes and Se-uptake by plants, a combined sorption-plant-box experiment was devised with *Oryza sativa* caryopses, 170 mL of Se-spiked nutrient solution between 0 and 10,000 µg/L Se as selenite or selenate and 8.5 g of kaolinite or goethite substrate. Plant-free sorption controls were prepared in glass bottles parallel to the plant-box experiments. In addition, single plants exposed to 500, 2000 and 10,000 µg/L Se as selenite or selenate were air-dried. Their speciation was determined by measuring XANES scans, while mapping of spatial Se distribution was achieved by using µXRF at ANKA, Karlsruhe.

Results for all plant Se-uptake showed a preferential partitioning of selenite to the root (70 % of total plant-Se) and selenate to the shoot (72 % of total plant-Se). Due to a lack of competing ions in a nutrient-free environment, plant Se treatment directly upon germination led to greater Se-uptake at lower Se concentration (0 - 1000 µg/L Se as selenite and 0 - 250 µg/L Se as selenate) compared to the nutrient solution uptake (maximum uptake: 177 mg/kg Se DW in roots and 75 mg/kg Se DW in shoots for selenite; and 367 mg/kg Se DW in shoots and 96 mg/kg Se DW in roots). However, this treatment induced phytotoxicity symptoms and strongly decreased Se uptake when these lower Se concentrations were exceeded. Treating the plants with nutrients and delaying Se exposure for 5 - 7 days led to greater plant tolerance for Se and steadily increasing Se-uptake with increasing applied Se concentration (159 mg/kg Se DW in shoots and 312 mg/kg Se DW in roots for the Se-addition of 2500 µg/L Se as selenite and 405 mg/kg Se DW in the shoots and 128 mg/kg Se DW in the roots for the Se-addition of 2500 µg/L Se as selenate). However, competition by nutrients reduced Se-uptake into the plants.

Based on µXRF data, Se was mainly present in the phloem and xylem of the plant, irrespective of selenite or selenate application. Se-speciation in the plant tissue depended on the species applied. Selenite treatment led to a mean of 93 % organic Se in roots and 72 % organic Se in shoots, the rest being selenite (7 % and 28 %, respectively). Selenate treatment led to a mean of 48 % organic Se, 16 % selenite and 36 % selenate in roots and 46 % organic Se, 15 % selenite and 39 % selenate in shoots.

In the sorption experiments, adsorption by selenite and selenate onto kaolinite and goethite was very different. Measurements of remaining Se concentrations in solution showed that selenite adsorption onto kaolinite and goethite was 61 and 99 %, respectively. Selenate's respective sorption was 72 and 42 %. Desorption showed that 71 - 96 % of previously adsorbed Se was exchangeable by ion exchange with phosphate. However, subsequent plant Se-uptake was not higher through contact with the mineral than it was in just the nutrient solution. This suggests that plants did not actively desorb Se from the mineral surface. It was also shown that the nutrient solution's effect on Se adsorption was complex and could not be described as a linear combination of the effects of single ions.

These experimental data were used to create a mass balance model of the Critical Zone. The model assumes 100 % of the Se to be distributed into the three compartments "mineral surface", "solution" and "plant". With the required input parameters limited to the concentrations of selenite and selenate, the model calculates the Se adsorption to the mineral surface, the resulting Se solution concentration and the subsequent Se concentrations in plant roots and shoots, as well as the Se speciation found in these tissues.

This model does not simply interpolate Se concentrations between experimental results, but can model the competition between selenite and selenate when both are present in solution, as well as calculate adsorption for less or more than the appointed kaolinite amounts used in the experiments. For each scenario, the complete mass balance is calculated. Unfortunately, goethite results could not be incorporated into the model as experimental Se concentrations did not approach maximum load capacity.

The model created here quantified key processes of Se transfer between mineral surface, solution and plant and can easily be extended using results of further experiments. This model, reproducing experimental results, is a solid basis for our ability to predict Se transfer in the Critical Zone. This is a crucial component of effective global Se resource management concerning biofortification and Se toxicity risk assessment.

Zusammenfassung

Selenressourcenverwaltung ist ein lebenswichtiges Thema. Dieses Halbmetall-Spurenelement ist für viele Organismen essentiell und weist eine geringe Spanne zwischen Mangelernährung (< 55 µg/Tag) und chronischer Toxizität (> 400 µg/Tag) für den Menschen auf. Überdies hinaus ist Se in der Welt ungleichmäßig verteilt und seine Bioverfügbarkeit durch viele Faktoren bestimmt. Biofortifikation und Selentoxizitäts-Risikoanalyse sind Schlüsselaspekte der globalen Selenressourcenverwaltung. Daher sind detaillierte Untersuchungen hinsichtlich molekularbiologischer und geochemischer Wechselwirkungen zwischen Se und der Umwelt notwendig. Dementsprechend war es Ziel dieser Studie, Selentransfer in der kritischen Zone, d.h. an der Schnittstelle zwischen Boden und Pflanze, zu beschreiben und zu quantifizieren.

Zu diesem Zweck wurde die kritische Zone in drei Selenreservoire unterteilt: "Boden", "Bodenlösung" und "Pflanze". Experimentell wurden diese durch die Modellminerale Kaolinit und Goethit, eine Nährlösung und Reis (*Oryza sativa*) repräsentiert. Unter der Verwendung identischer Versuchsparameter wurde zunächst der speziesabhängige Selentransfer von der Lösung in die Pflanze und die darauffolgende Selenpartitionierung innerhalb der Pflanze untersucht. Dann wurden Adsorptions-Desorptionsprozesse von Selenat und Selenit an den Modellmineralen Kaolinit und Goethit erforscht. Danach wurden beide Experimente kombiniert. Die Quantifizierung des Selentransfers zwischen Lösung und Mineralsubstrat sowie Selenaufnahme von der Lösung in die Pflanze stellen die Basis für ein mathematisches Prozessmodell dar.

Um Unterschiede in der Selenaufnahme aufgrund von Nährstoffmangel zu bestimmen, wurde eine Serie an geschlossenen Pflanzenboxversuchen genutzt. *Oryza sativa*-Karyopsen wurden mit einem mit Selen versetztem, nährstofffreiem Agar oder einer Nährlösung von 0 - 2500 µg/L Se als Selenit oder Selenat vorbereitet. Pflanzen zu zwingen, durch ein Se-freies, Agar-gefülltes Eppendorfgefäß zu wachsen, bevor sie mit der mit Selen versetzten Nährlösung oder Phytoagar in Kontakt kommen, ermöglichte es darüber hinaus, die Effekte verzögerter Selenbehandlung zu beobachten. Alle Pflanzen wurden mit 14-tägiger Selenaufnahme in einer Klimakammer kultiviert. Nach der Ernte wurden Pflanzensammelproben des Sprosses und der Wurzel gefriergetrocknet und mit Salpetersäure und Wasserstoffperoxid in der Mikrowelle aufgeschlossen. Selengehalte wurden mit der HG-FIAS bestimmt. Agarproben wurden ähnlich aufgeschlossen und mit der HG-FIAS gemessen. Selen- und Nährstoffgehalte in der Nährlösung wurden mit ICP-MS und IC gemessen.

Um das Selenadsorptions- und -desorptionsverhalten zu charakterisieren, wurde eine Serie von 24-h Batchsorptionsversuchen mit 0,5 g Kaolinit oder Goethit in einer der Nährlösung entsprechenden KCl-Ionenstärke durchgeführt. Die Lösungen waren mit 0 - 5000 µg/L Se in Form von Selenit oder Selenat versetzt. Um Konkurrenzeffekte zu untersuchen, wurden diese Versuche in Anwesenheit von 750 µmol/L Nitrat, Sulfat oder Phosphat wiederholt. Um das Ionenaustauschpotential zu quantifizieren, wurden anschließend Selendesorptionsversuche mit 0,1 mol/L K_2HPO_4 durchgeführt.

Um Kombinationseffekte zwischen Selensorptionsprozessen und Pflanzenselenaufnahme zu untersuchen, wurde ein kombiniertes Sorptions-Pflanzenbox-Experiment konzipiert, mit *Oryza sativa*-Karyopsen,

170 mL mit Selen versetzter Nährlösung zwischen 0 und 10.000 µg/L Se in Form von Selenit oder Selenat und 8,5 g Kaolinit- oder Goethitsubstrat. Parallel zu den Pflanzenboxexperimenten, wurden pflanzenlose Sorptionskontrollen in Glasflaschen angesetzt. Zusätzlich wurden einzelne Pflanzen, die 500, 2000 und 10.000 µg/L Se in Form von Selenit oder Selenat ausgesetzt waren, luftgetrocknet. Deren Speziation wurde durch XANES Messungen bestimmt und die Kartierung der räumlichen Selenverteilung wurde durch Verwendung von µXRF an der ANKA, Karlsruhe, erreicht.

Die Ergebnisse für die Selenaufnahme aller Pflanzen zeigte eine bevorzugte Partitionierung des Selenits in die Wurzel (70 % des gesamten Pflanzenselens) und des Selenats in den Spross (72 % des gesamten Pflanzenselens). Aufgrund fehlender Konkurrenzionen in der nährstofffreien Umgebung, führte die direkt zum Zeitpunkt der Keimung erfolgte Selenbehandlung zu größerer Selenaufnahme bei niedrigerer Selenkonzentration (0 - 1000 µg/L Se als Selenit and 0 - 250 µg/L Se als Selenat) im Vergleich zur Aufnahme in der Nährlösung (maximale Aufnahme: 177 mg/kg Se DW in der Wurzel und 5 mg/kg Se DW im Spross für Selenit, und 367 mg/kg Se DW im Spross und 96 mg/kg Se DW in der Wurzel). Diese Behandlung erzeugte jedoch Phytotoxizitätserscheinungen und stark verringerte Selenaufnahme bei Überschreitung dieser geringen Selenkonzentrationen. Pflanzenbehandlung mit Nährstoffen und einer um 5 - 7 Tage verzögerten Selenexposition führte zu größerer Selentoleranz bei Pflanzen und stetig ansteigender Selenaufnahme mit steigender zugegebener Selenkonzentration (159 mg/kg Se DW im Spross und 312 mg/kg Se DW in der Wurzel für die Zugabe von 2500 µg/L Se als Selenit und 405 mg/kg Se DW im Spross und 128 mg/kg Se DW in der Wurzel für die Selenzugabe von 2500 µg/L Se als Selenat).

Konkurrenz durch Nährstoffe verringerte jedoch die Selenaufnahme in die Pflanze.

Basierend auf µXRF-Daten, war Se vor allem im Phloem und Xylem vorhanden, unabhängig von der Selenzugabe in Form von Selenit oder Selenat. Die Selenspeziation im Pflanzengewebe war von der zugegebenen Spezies abhängig. Selenitbehandlung führte zu einem Mittelwert von 93 % organischem Se in der Wurzel und 72 % organischem Se im Spross, während der Rest aus Selenit bestand (7 % und 28 %, respektive). Selenatbehandlung führte zu einem Mittelwert von 48 % organischem Se, 16 % Selenit und 36 % Selenat in der Wurzel und 46 % organischem Se, 15 % Selenit and 39 % Selenat im Spross.

In den Sorptionsversuchen war Adsorption von Selenit und Selenat an Kaolinit und Goethit sehr unterschiedlich. Messungen von Selenrestkonzentrationen in der Lösung zeigten eine Selenitadsorption an Kaolinit und Goethite von 61, bzw. 99 %. Die Selenatsorption lag bei 72 bzw. 42 %. Desorption zeigte, dass 71 - 96 % des zuvor adsorbierten Se's durch Ionenaustausch mit Phosphat austauschbar war. Die darauffolgende Pflanzenselenaufnahme war durch den Kontakt mit dem Mineral jedoch nicht höher als in der reinen Nährlösung. Dies lässt vermuten, das Pflanzen Se nicht aktiv von der Mineraloberfläche desorbiert haben. Es wurde auch gezeigt, dass der Effekt der Nährlösung auf die Adsorption komplex ist und sich nicht als eine Linearkombination der Effekte einzelner Ionen darstellen lässt.

Diese experimentellen Daten wurden verwendet, um ein Massenbilanzmodell der kritischen Zone zu erstellen. Das Modell nimmt an, dass 100 % des Se's auf die drei Kompartimente "Boden", "Bodenlösung" und "Pflanze"verteilt sind. Mit den auf die Selenit- und Selenatkonzentration beschränkten notwendigen Eingabeparameter berechnet das Modell die Selenadsorption an der Mineraloberfläche, die resultierende Selenkonzentration in der Lösung und die darauffolgenden Selenkonzentrationen in Pflanzenwurzeln und -sprossen sowie die in diesen Geweben vorhandene Selenspeziation.

Dieses Modell interpoliert nicht einfach zwischen Selenkonzentrationen experimenteller Ergebnisse, sondern kann die Konkurrenz zwischen Selenit und Selenat modellieren, wenn beide in der Lösung vorhanden sind und kann die Adsorption für geringere oder größere als in den Experimenten verwendete Mengen an Kaolinit berechnen. Für jedes Szenario wird eine vollständige Massenbilanz berechnet. Unglücklicherweise konnten Goethitergebnisse nicht in das Modell integriert werden, da bei den experimentellen Selenkonzentrationen die maximale Beladungskapazität nicht annahernd erreicht wurde.

Das Modell, dass hier erstellt wurde, quantifiziert Schlüsselprozesse des Selentransfers zwischen Mineraloberfläche, Lösung und Pflanze und kann leicht durch die Einbindung weiterer Versuche erweitert werden. Dieses Modell, welches experimentell erzielte Ergebnisse reproduziert, ist eine solide Basis für die Prognose des Selentransfers in der kritischen Zone. Dies stellt eine entscheidende Komponente effektiver globaler Selenressourcenverwaltung im Hinblick auf Biofortifikation und Selentoxizitäts-Risikoanalyse dar.

"I do not know what I may appear to the world, but to myself I seem to have been only like a boy playing on the sea-shore, and diverting myself in now and then finding a smoother pebble or a prettier shell than ordinary, whilst the great ocean of truth lay all undiscovered before me."

–Sir Isaac Newton–

Table of Contents

List of Figures

List of Tables

1 Introduction

The story of Selenium (Se) research is one littered with serendipity [Oldfield, 2006]. It began with the accidental discovery of Se nearly 200 years ago by the Swedish chemist Jöns Jacob Berzelius [Berzelius, 1817] and continued throughout the next two centuries until the beneficial effects of Se were discovered while studying ways to avoid its toxicity in the second half of the 20th century [Oldfield, 2006]. To this day, new selenoproteins are still being discovered and not all metabolic functions and geochemical pathways of many organic species and some inorganic species have been clarified [Khan & Hell, 2014]. The transformation that the view on Se has undergone throughout the past two centuries is indicative of the surprises and challenges scientists are faced with when studying this trace element.

1.1 The fate of Se in research

For decades, Se had been considered to be merely a toxic trace element, which was discovered to cause hair and hoof loss in animals during the 1930s [Lenz & Lens, 2009]. It wasn't until 1957 that Se deficiency was recognized as a potential cause for human disease [Schwartz & Foltz, 1957].

In 1973, it was determined that, indeed, this trace element was interesting not only due to its toxicity [Oldfield, 2006], but also because it was proven to be essential to humans, as it is an integral component of the glutathione peroxidase enzyme. This antioxidant compound is designed to protect hemoglobin from oxidative damage [Flohé et al., 1973, Rotruck et al., 1973].

This was only the first of various seleno-proteins to be discovered in the human body which, in combination, account for a wide array of functions within the human organism [Rotruck et al., 1973]. Since then, Se has been acknowledged as having 'two faces' [Oldfield, 1987] labelled 'the essential toxin' [Lenz & Lens, 2009] or 'a double edged sword' [Fernández-Martínez & Charlet, 2009, Hartikainen, 2005] to describe its dual nature.

These two facets of Se had been known to scientists for years. However, awareness of Se as an environmental pollutant became acute in the Kesterson Reservoir controversy in the United States in the 1980s not only for scientists, but also for politicians and the general public [Terry et al., 2000]. To reduce salt build-up, agricultural drainage water, which also happened to be geogenically enriched in Se, was discharged into the reservoir. When reproductory failure, development defects and mortality were found in migratory aquatic birds and fish, these were linked to the anthropogenic activity [Terry et al., 2000, Winkel et al., 2011].

This is not the only form of anthropogenic Se-contamination of the environment; industrial activities such as oil refineries and electric utilities also generate Se-contaminated aquatic discharges [Terry et al., 2000]. Amelioration and management of such sites still present considerable research potential today [Water Education Foundation, 2014].

On the other hand, Se is considered a rare and non-renewable resource, which must be carefully managed [Haug et al., 2007]. Optimizing procedures of crop fertilization of Se-deficient areas, such as Finland, still shows great potential for improvement [Eurola et al., 2003, Winkel et al., 2011]. Therefore, Se research faces two equally important, yet entirely diverse goals:

1. securing Se nutrient resources for future generations and
2. management of current enormous Se-enriched waste deposits to protect the environment and improve the quality of life in such areas of contamination [Haug et al., 2007].

Understanding the underlying geochemical and biochemical mechanisms for Se transfer processes within the Critical Zone is, therefore, an important first step for both aspects of these vital research questions.

1.2 Study objective and approach

This study is based on the premise that due to the complexity of Se geochemistry, a much better understanding of the Se cycle, as well as the basic processes governing it, is required [Fernández-Martínez & Charlet, 2009]. The objective of this study is, therefore, to quantify and characterize Se transfer pathways and processes as part of the Se cycle within the Critical Zone. To achieve this, the concept of this study relies on the compartmentalization of the Critical Zone into three Se reservoirs: 'soil', 'soil solution' and 'plant' (Figure 1.1). With the reservoir 'soil solution' connecting the reservoirs 'soil' and 'plant', this enables the study of two main process pathways for Se cycling on a scale of laboratory experiments:

1. speciation- and concentration-dependent uptake of Se into plants using the model plant of *Oryza sativa* in the absence or presence of nutrients in agar medium or a solution with defined chemical characteristics, respectively;

2. Se sorption processes between soil particles and the soil solution using the two model minerals of kaolinite and goethite with the focus on Se ad- and desorption processes in the presence of the competing plant-relevant oxy-anions nitrate, sulfate and phosphate.

The combination of both of these Se transfer processes and pathways are then connected to produce an empirical process model of the biogeochemistry of the Se cycle between these compartments of the Critical Zone.

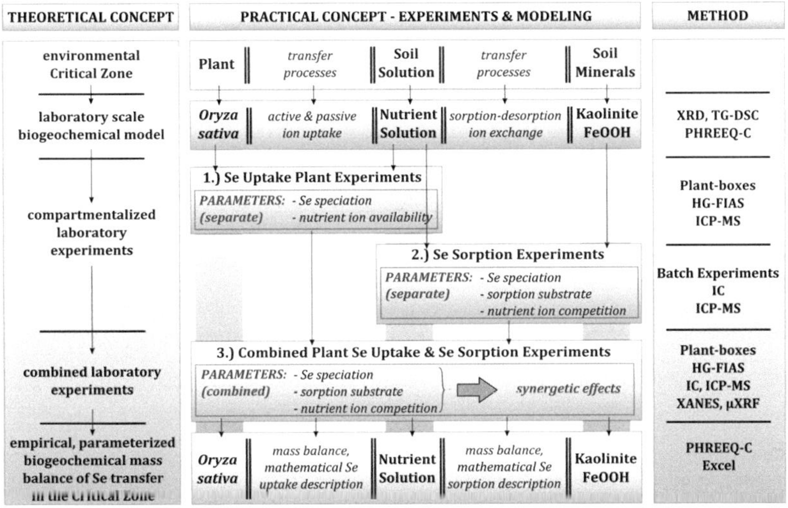

Figure 1.1: Research Concept: Quantification and characterization of processes between three compartments of the Critical Zone: soil, soil solution and plant. A 3-step experimental approach works toward
(1) understanding Se pathways between solution and plants depending on Se speciation, nutrient availability and nutrient competition (explored by experiments A1 - A4),
(2) understanding influences on Se pathways from sorption processes depending on Se speciation and ion competition (explored by experiments B1 - B3) and
(3) creating a biogeochemical process model accounting for synergistic effects between both experiments (explored by experiment C).

2 Selenium in the environment

Selenium was discovered in 1817 due to its geochemical similarity to sulfur. It was Jöns Jacob Berzelius, who, in the course of studying methods to prepare sulfuric acid from sulfur-bearing rock, came across a reddish sludge that formed when using pyrite from the mines in Falun, Sweden. This sludge contained an element that could be identified neither as sulfur, arsenic nor tellurium [Berzelius, 1817, Fernández-Martínez & Charlet, 2009].

2.1 Geochemical properties of Se

Selenium is a nonmetal trace element with the atomic number 34 and has an atomic mass of 78.96 g/mol [Fernández-Martínez & Charlet, 2009]. Like oxygen (O) and sulfur (S), it belongs to number VI of the main groups of the periodic table of elements and has an electron configuration of [Ar] $3d^{10}$ $4s^2$ $4p^4$. Similar to S, this leads to the five natural oxidation states of Se: -II, -I, 0, +IV, +VI [Neal, 1995]. Selenium's two oxy-anions, which are the focus of this study, are the fully oxidized tetrahedral selenate and the pyramidal selenite.

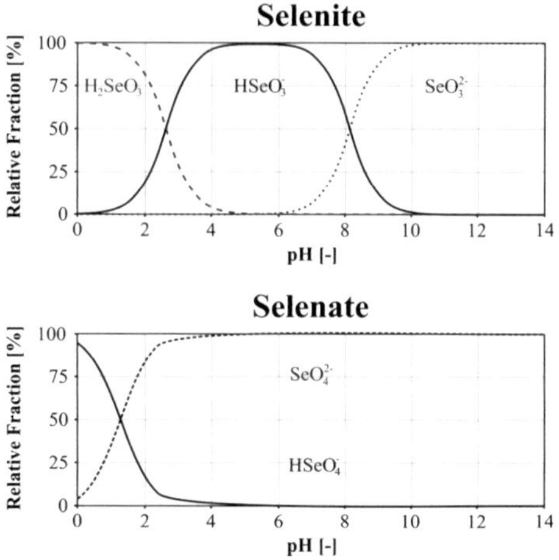

(a)

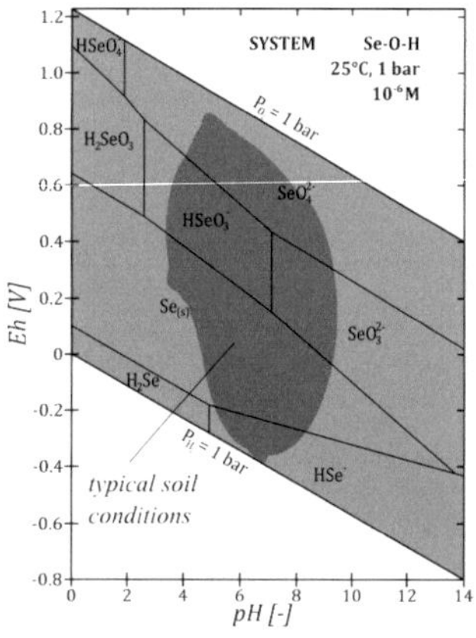

(b)

Figure 2.1: pH-dependent Se species
a: Bjerrum plot for selenite and selenate: c(Se) = 0.01 mol/L, T = 25 °C, p = 1 bar (created with PHREEQ-C, wateq4)
b: Pourbaix diagram of the Se-H-O system: c(Se) = 10^{-6} mol/L, T = 25 °C, p = 1 bar [Brookins, 1988]

Similar to sulfuric acid, selenic acid (H_2SeO_4) is a strong acid [Mortimer, 2010]. In aqueous solution, selenate (SeO_4^{2-}) and the once-protonated biselenate ($HSeO_4^-$) exist with a dissociation constant of pK_{a2} = 1.80 ±0.10 (Figure 2.1a). The doubly protonated species (H_2SeO_4) has a constant of pK_{a1} = -2.01 ±0.06 and, therefore, does not exist under natural conditions [Séby et al., 2001].

Selenous acid (H_2SeO_3), on the other hand, is a weak acid, analogous to sulfurous acid [Mortimer, 2010]. Selenite exists in solution as H_2SeO_3, $HSeO_3^-$ and SeO_3^{2-} with dissociation constants of pK_{a1} = 2.70 ±0.1 and pK_{a2} = 8.54 ±0.04, depending on the solution pH [Séby et al., 2001] (Figure 2.1a).

As shown in Figure 2.1b, both Se oxy-anions are dominant under oxic conditions. The less oxic conditions become, however, the more elemental selenium, Se(0), is found to be prevalent (Figure 2.1b). Eleven Se(0) allotropes with various structural properties have been described [Minaev et al., 2005], all of which are believed to be present in nature in non-soluble form and, therefore, of limited bioavailability [Fernández-Martínez & Charlet, 2009]. However, recent studies have also shown plausible evidence of biological availability of Se(0) nanoparticles [Terry et al., 2000, Zhang et al., 2005].

Under strongly reducing conditions, Se(-II) and Se(-I) are stable (Figure 2.1b), generally as a variety of metallic selenides and organic compounds [Fernández-Martínez & Charlet, 2009]. Also present under strongly reducing conditions is the gaseous, inorganic compound H_2Se (Figure 2.1b), which has been shown to be a product of microbiological processes [Fernández-Martínez & Charlet, 2009].

There are five stable Se isotopes (given in order of their abundance): ^{80}Se (49.61 %), ^{78}Se (23.78 %), ^{76}Se (9.36 %), ^{77}Se (7.63 %), ^{74}Se (0.87 %). There are also multiple radioactive isotopes as well, one of which is the naturally occurring ^{82}Se which, due to its long half-life of $1.1 \cdot 10^{20}$ years, makes up 8.73 % of naturally occurring Se isotopes [Boullis, 1997, Duc et al., 2003]. Therefore, research on Se behavior in the environment is also applicable to the mobility of radioactive Se, which is of interest for long-term nuclear waste disposal [Bitterli et al., 2010].

As both are members of the VI-A chalcogen group of elements, Se and S share many properties. This means that Se can replace S in compounds, minerals and proteins. However, the Se atom has a larger radius (0.5 $\AA$) than the S atom (0.37 $\AA$), which makes a Se-Se bond approximately 1/7 longer and 1/5 weaker than a S-S bond [Sors et al., 2005].

2.2 Abundance and anthropogenic use of Se

The primary source of Se in the terrestrial system is rock [Rosenfeld & Beath, 1964, Neal, 1995]. Forty percent of the Se found in the earth's crust is bound in rocks, mainly as clay fractions in sediments, shales (Table 2.1), phosphatic rocks, coals or organic-rich deposits that contain high concentrations of Se, while sedimentary rocks are usually low in Se [Fernández-Martínez & Charlet, 2009].

Selenium abundance in the earth's crust is around 0.05 mg/kg, which equals about 1/6000 of the total S content [Wang & Gao, 2001, Haug et al., 2007]. However, it is very unevenly distributed throughout the world, ranging from concentrations near zero up to 1250 mg/kg [Haug et al., 2007].

Table 2.1: Se-contents in the environment, from [Fernández-Martínez & Charlet, 2009]

Compartment	**Se [mg/kg]**	**Compartment**	**Se [mg/kg]**
Earth's Crust	0.05	Soil	0.01 - 2
Igneous & Volcanic Rock	0.35	Se-deficient Soils (China)	0.004 - 0.48
Limestone	0.03 - 0.08	Atmospheric Dust	0.05 - 10
Carbon Shale (China)	206 - 280	Freshwater [µg/L]	0.02

Selenium is often found substituting S in minerals such as chalcopyrite, pyrite or other sulfides or substituting phosphorus in phosphate minerals [Fernández-Martínez & Charlet, 2009]. However, Se-bearing minerals are too scarce to be mined [Haug et al., 2007]. Therefore, Se is produced as a by-product during electrolytic refining of copper, lead and silver or as a by-product in the manufacturing of sulfuric acid [Adriano, 1986, Stwertka, 2002].

Selenium also occurs as a by-product of radioactive decay; ^{79}Se, a long-living Se isotope has an abundance of 0.04 % in nuclear waste disposal and a relatively long half-life of $6.5 \cdot 10^4$ years, [Boullis, 1997, Duc et al., 2003, Bitterli et al., 2010].

Selenium is utilized for many purposes, such as the manufacturing of glass, plastics, ceramics and electronic components (because of its properties as a semi-conductor). It is also used in the chemical industry in pigments and lubricants or in shampoos against dandruff and also in nutritional supplements in combination with vitamin E [Adriano, 1986, Neal, 1995, Haug et al., 2007].

Because soil Se concentration is mainly influenced by the parent rock, the average world mean Se content in soils is 0.4 mg/kg (0.01 - 2 mg/kg). Soils with a Se content below 0.1 mg/kg are considered Se-deficient [Dhillon & Dhillon, 2003]. Soils above 0.5 mg/kg of Se are considered seleniferous because, in the absence of knowledge about their Se speciation, this Se concentration poses the risk of producing forage which exceeds maximum advisable Se-levels for animal consumption [Dhillon & Dhillon, 2003]. Both types of soil can exist in close proximity of each other (less than 20 km apart), presenting some countries, such as China and Brazil, with both challenges associated with this double-edged sword element [Fordyce, 2007].

As shown in Figure 2.1, Se in aerobic and neutral to alkaline environments tends to be present as selenate and selenite, whereas the major species dominating anaerobic environments are selenide and elemental Se [Adriano, 1986, Terry et al., 2000]. Volatile Se compounds are also estimated to constitute a large part of the Se cycle with 13,000 to 19,000 t cycling through the troposphere every year [Wen & Carignan, 2007] as dimethylselenide (DMSe), dimethyldiselenide (DMDSe) and probably dimethyl selenone, dimethyl selenylsulfide and methaneselenol [Reamer & Zoller, 1980].

Comparing Se content in the environmental compartments listed in Table 2.1, coupled with knowledge of their abundance, it is estimated that Se-deficient environments are more widespread than Se-abundant ones [Haygarth, 1994].

2.3 Selenium in human health – requirement and toxicity

On the one hand, Se has been proven to be essential for mammals, and therefore, humans. In a wide array of 40 seleno-proteins known to be found in mammals, this trace element not only acts as an important anti-oxidant [Oldfield, 2006]. It is known to act as a growth factor as well, playing an important role in catalyzing production of thyroid hormones and in synthesising leucotrienes, which are needed to regulate immune responses. Furthermore, Se is involved in regulatory processes of a multitude of other metabolic functions [Reilly, 1997, Imai, 1998, Rayman, 2000]. To this day, an increasing number of seleno-proteins are still being discovered [Khan & Hell, 2014].

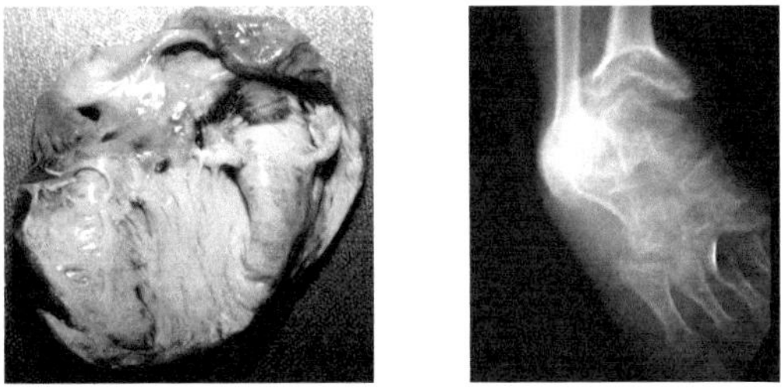

(a) (b)

Figure 2.2: Pathological Se deficiency
a: necrosis and mineralization in heart muscle tissue, [Beytut et al., 2002];
b: X-ray image of the left ankle of a 14-year old Tibeten boy with Kashin-Beck disease showing deformed bones and ligaments, [Moreno-Reyes et al., 1998]

Suboptimal supply of Se is, therefore, associated with a number of adverse health effects such as myalgia, cartilage dysfunction, oxidative stress, heart failure, impaired immune function, reduced fertility and an elevated risk of certain cancers [Combs, 2001, Rayman, 2008]. Currently, an estimated 0.5 - 1 billion people are believed to be insufficiently supplied with Se [Combs, 2001]. Its essential role as a micronutrient becomes very clear in the effects of extreme Se deficiency: as shown in Figure 2.2a, low Se intake is known to lead to 'white muscle disease' in cattle [Oldfield, 2006], which is characterized by white necrosis and mineralization of calcium deposits within the heart muscle [Beytut et al., 2002]. An illness related to Se deficiency in humans is Kashin-Beck disease [Jiyun et al., 1982, Levander & Burk, 2006], which is characterized by distrophic osteoarthrosis in its first stage (Figure 2.2b). On reaching the second stage, this becomes chronically painful and leads to irreversible deformities in its third stage [Nesterov, 1964].

On the other hand, Se toxicity can reach chronic or even acute poisoning levels. The so-called 'alkali disease' in livestock is attributed to chronic Se poisoning (selenosis), demonstrating symptoms such as hair loss, hoof deformity, anemia and the wasting away of the body [Lombeck et al., 1987]. Acute Se poisoning is sometimes equated to 'blind staggers', another disease described in cattle, the symptoms of which include blindness, abdominal pain, paralysis and death through loss of appetite [Lombeck et al., 1987]. However, contrary to assumptions made in the 1930s and '40s, the nosology of 'blind staggers' could not be clearly linked to Se poisoning [O'Toole & Raisbeck, 1964].

Se-related illnesses in humans are more likely to be induced by Se deficiency, although there are naturally seleniferous regions in China or Venzuela whose inhabitants have been affected by chronic selenosis [Lombeck et al., 1987]. Cases of acute Se poisoning are rare; however, there also exists a published case study of acute Se poisoning in a 2-year old child [Lombeck et al., 1987]. Toxicity symptoms of hypochromic anaemia, leucopoenia and damaged nails were reported in workers with prolonged Se exposure in manufacturing, while accidental ingestion of high Se concentrations has been related to vomiting, diarrhoea and neurological disturbances [Navarro-Alarcon & Cabrera-Vique, 2008].

It wasn't until 1996 [Levander & Burk, 2006] that the World Health Organization (WHO) introduced references on Se intake, after Chinese studies were able to provide data on recommended daily allowance (RDA), accounting both for the dietary need of Se as an anti-oxidant and its adverse effects when consumed in toxic concentrations. Today's recommended Se intake – according to the 2000 Dietary Reference Intakes (DRI) – advises 20 - 30 µg Se per day for children up to the age of 8 and 55 µg of Se per day for adults, both male and female. 400 µg of Se per day is considered to be the upper tolerance limit of daily intake [Institute of Medicine, 2000, Levander & Burk, 2006].

This range between dietary deficiency (< 55 µg per day) and toxicity (> 400 µg per day) is one of the narrowest that can be found for any element [Finley, 2005, Fernández-Martínez & Charlet, 2009]. Some authors consider this range of health benefits to be even smaller at 55 - 200 µg Se per day [Rayman, 2000].

2.4 Biochemical properties of Se

Selenium can be taken up by organisms as selenate, selenite, organic Se compounds, such as selenomethionine (SeMet) and also as Se(0) nanoparticles [Terry et al., 2000, Fernández-Martínez & Charlet, 2009].

> *"The bioavailability of a trace element is related to factors that make this element available to an organism, in a form that is transportable across the organisms' biological membrane."*
> *[Reeder et al., 2006, Fernández-Martínez & Charlet, 2009]*

Factors of bioavailability relate to the solubility of interacting substances, which are affected by speciation, ionic strength, pH or redox potential [Fernández-Martínez & Charlet, 2009]. These, in turn, are the result of indirect factors, such as soil type, plant species and amounts of rainfall [Girling, 1984, Haug et al., 2007]. Therefore, bioavailability is ultimately dependent on bedrock geology and climate as is apparent in the case of Se-deficient Scandinavian soils of Norway and Finland [Xue et al., 2001]. Due to the low bedrock-Se coupled with low temperatures, high humidity and low pH as well as a high Fe content, any Se in these soils of glacial origin is present in reduced and therefore bio-unavailable form [Oliver, 1997]. Furthermore, not only soil development, but also processes of soil erosion have an impact on soil-Se bioavailability and play a key role in Se soil deficiency, as evident in Chinese soils [Oliver, 1997].

With so many factors governing Se transport and bioavailability, it comes as no surprise that total Se soil concentrations do not necessarily reflect whether plant uptake of Se will cause toxicity or deficiency [Lakin, 1972]. This explains the seemingly contradictory observation that 0.03 mg/kg Se in soils of Kars, Turkey lead to an accumulation of Se in meadow grass of 0.07 mg/kg Se dry mass on average, which then resulted in white muscle disease in cattle [Beytut et al., 2002], while in parts of Finland, 0.2 mg/kg Se soil, nearly a tenfold total concentration compared to that of the Kars region, is considered deficient in Se due to the element's reduced form, which is not readily taken up by plants [Oliver, 1997].

While organic Se as selenide (Se(-II)) is considered the most bioavailable Se form in marine environments, as it is taken up 1000 times more readily than inorganic Se by algae [Lemly, 1993], in terrestrial ecosystems, selenate (Se(VI)) is considered the most mobile and therefore most bioavailable form of Se [Haug et al., 2007]. Selenite (Se(IV)), though potentially 5 - 10 times more bioavailable and toxic than selenate [Lemly, 1993], is often retained in the soil due to fixation by soil minerals, such as ferric oxides or organic matter [Haug et al., 2007]. Elemental Se(0), on the other hand, is insoluble and therefore considered to have little toxicological significance unless perhaps as nano-particles [Fernández-Martínez & Charlet, 2009, Winkel et al., 2011].

Microorganisms are known to perform biotic transformations of Se. Organic Se compounds, selenate and selenite can be reduced actively by bacteria in the process of dissimilatory reduction (using the anion as a terminal electron acceptor) or assimilatory reduction (incorporating Se into organic compounds) to selenite, elemental Se and selenide [Fernández-Martínez & Charlet, 2009]. There are at least 16 species of bacteria or archaea known to grow under anaerobic conditions by linking the oxidation of organic substrates or H_2 to the dissimilatory reduction of selenium oxy-anions [Fernández-Martínez & Charlet, 2009].

The environmentally most important process in this regard is dissimilatory reduction [Stolz & Oremland]. Thermodynamic data show that the free energy for SeO_4^{2-} and SeO_3^{2-} reduction into selenide (HSe^-) is -22.2 and -22.9 kcal eq^{-1}, respectively, which is about four times higher than that for the reduction of SO_4^{2-} to sulphide (HS^-; -5.9 kcal eq^{-1}). Therefore, selenate reduction to selenite is favourable over nitrate reduction to ammonium and sulfate reduction to sulfide, which makes selenate an environmentally significant electron acceptor [Stolz & Oremland]. Selenate reduction has been found to occur in chemically disparate environments [Stolz & Oremland] and in many cases, selenite has been shown to be an intermediate product of this reduction, which is believed to subsequently adsorb onto soil minerals [Fernández-Martínez & Charlet, 2009]. However, it has also been shown that microbiological selenite reduction occurs at a faster rate than selenate reduction [Bajaj et al., 2011] which suggests rapid metabolization in cases in which it is not rapidly adsorbed.

The ability of micro-organisms to perform bio-methylation of Se is often seen as a possible remediation procedure for Se-contaminated soils. This process consists of a series of redox reactions changing the Se oxidation state from +IV or +VI to -II, by producing selenide compounds with methyl groups [Chasteen & Bentley, 2003]. These species can be volatile and are, therefore, removed from the soil. This is important because although much of plant Se-uptake occurs via soil or solution, it has also been shown that even volatile Se species can be taken up by plants via absorbtion through the leaf [Zieve & Peterson, 1984, Xu & Hu, 2004].

Plants are believed to take up selenate via the sulfate transporter, while uptake of selenite by sulfate, phophate or silicone transporters is still debated [Terry et al., 2000, Li et al., 2007, Zhao et al., 2010]. Selenium can then be transformed into organic Se compounds and also volitalized [Terry et al., 2000]. Details on the uptake of Se into plants and translocation within plants are provided in Chapter 3.1.

In many vertebrates, Se is known to replace S in proteins due to similarities between these two elements, most frequently in seleno-cysteine (SeCys) compared to cysteine (Cys). In direct comparison between enzymes containing SeCys rather than Cys, the seleno-isologues prove to be the more efficient catalysts [Birringer et al., 2002]. In the case of glutathione peroxidase (GSH), which is required by all mammals, efficiency of this enzyme for detoxifying hydroperoxides is reduced by 2 - 3 orders of magnitude when Se does not substitute S and, therefore, fails to fulfill its purpose. The reason that Se is essential to organisms is, therefore, attributed to this substitution of S in proteins with Se, which, leading to greater reactivity than in their S-counterpart, is often required for catalytic activity [Birringer et al., 2002, Sors et al., 2005].

In general, biomolecular mechanisms of Se toxicity are poorly understood [Brozmanová et al., 2010]. However, the same greater reactivity of Se-substituted proteins may also be the cause for its toxicity. As an oxidizing catalyst, Se is able to oxidize thiols producing free radicals which can damage DNA. Moreover, unwanted substitutions of S by Se in proteins leads to disfunctional proteins, causing hoof, hair, nail and feather loss when S of the keratin is substituted for Se [Spallholz & Hoffman, 2002]. SeCys provides the greatest source of Se-toxicity to humans as it leads to accumulation of hydrogen selenide due to inhibition selenium methylation [Spallholz & Hoffman, 2002].

2.5 Approaches to studying Se in the environment

Introduction of Se into the foodchain is a source of both essential nutritional supply to organisms as well as environmental toxicity hazardous to life [Lenz & Lens, 2009, Winkel et al., 2011]. Because of the narrow range between dietary deficiency and toxicity [Fernández-Martínez & Charlet, 2009], management thereof requires detailed understanding of Se transfer processes in all environments that include organisms. This concept is known as the 'Critical Zone'.

> *"The Critical Zone is the system of coupled chemical, biological, physical, and geological processes operating together to support life at the Earth's surface. (It) ... is defined as the volume extending from the upper limit of vegetation down to the lower limit of groundwater." [Anderson et al., 2007, Brantley et al., 2007]*

This geo-science model concept includes the multitude of natural components that are interwoven with living organisms and also coincides with compartments of an element's cycle: the geosphere, pedosphere, biosphere, hydrosphere and atmosphere.

The challenge that presents itself in the environmental modelling of Se is the amount of parameters governing Se bioavailability, transfer and retention. Furthermore, Se is known to behave non-conservatively in the environment. Moreover, it cannot simply be described by applying transport pathways of sulfur [Presser & Piper, 1998]. To tackle this problem, there have been various approaches to studying Se in the environment (Figure 2.3). Selenium transfer between compartments of the Critical Zone can be studied on various scales and modelling of Se in the Critical Zone can be divided accordingly into different groups (Figure 2.3).

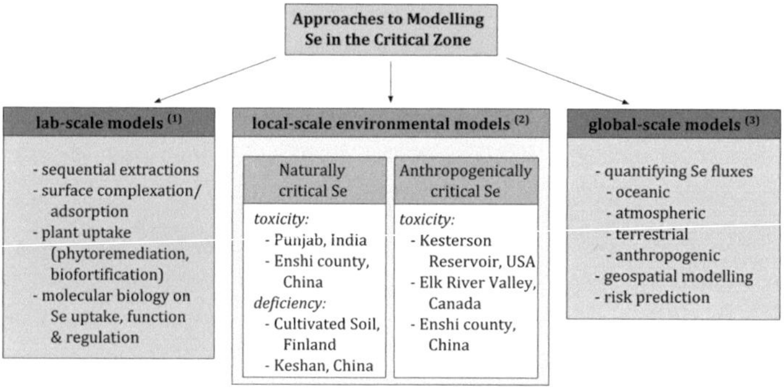

Figure 2.3: Approaches to modelling Se in the Critical Zone on different scales: (1) theoretical and empirical lab-scale approaches [Balistrieri & Chao, 1990, Kulp & Pratt, 2004, Li et al., 2007, Khan & Hell, 2014], (2) empirical local-scale approaches [Presser & Piper, 1998, Wang & Gao, 2001, Ohlendorf, 2002, Bajaj et al., 2011], (3) and theoretical, empirical or spatial-numerical global-scale approaches [Haygarth, 1994, Bowie et al., 1996, Amoroux, 2001, Winkel et al., 2011]

Although there had been previous studies on lab-scale effects of Se [Lévesque, 1974, Asher et al., 1977, Howard, 1977, Tzeng & Zeitlin, 1978], it was the Kesterson Reservoir controversy in the 1980s that brought awareness of Se as an environmental pollutant to the broader public [Terry et al., 2000], thus paving the way for more interdisciplinary, holistic approaches to Se research on a larger scale. Subsequent studies conducted on the deaths of Kesterson waterfowl and other birds have been recognized as one of the 'gold standards' of retrospective ecological risk assessment [Ohlendorf, 2002] with their integrated combination of field and laboratory studies that included recognising a disease syndrome, identifying the causative agent and verifying the findings in reproducing the disease syndrome in healthy individuals. It was the understanding of the complex interactions between anthropogenic effects – such as the re-routing of drainage water – in combination with geological effects – such as the naturally Se-enriched Cretacious marine source bedrock – which gave rise to the term "Kesterson Effect" [Presser, 1994, Water Education Foundation, 2014]. This term summarizes all aspects found to contribute to the biological accumulation of Se-induced toxicity, tracing Se "from rock to duck" [Presser, 1994, Water Education Foundation, 2014].

Ideally, all three scales (lab-scale, local-scale and global-scale, Figure 2.3) of Se transfer processes merge into a coherent model of the Se cycle, supporting both theoretical and empirical study. Within the scope of this study, the goal is to merge theoretical and practical lab-scale processes at an interdisciplinary level, combining a botanical and geochemical approach.

Small lab-scale systems (Figure 2.3(1)) have the advantage of having a more limited amount of parameters and being more clearly definable than larger, environmental-scale systems. All the more so, when the study can be conducted as a closed system. However, because of this restricted amount of influencing parameters, applicability of such small-scale systems is limited in the context of modelling the Se cycle. This is especially true of mono-disciplinary approaches.

For example, results from lab-scale studies of interactions between mineral surface and solution, such as adsorption/desorption studies, are very selectively applicable only to the mineral, pH-value and ionic strength chosen in the experiment [Parks, 1990]. And on the other hand, lab-scale bio-molecular approaches on the uptake of Se by plants and microorganisms target specific enzymatic processes of Se uptake, transport and assimilation [Terry et al., 2000]. Each of these important findings, however, must be combined into a coherent model when modelling the Se cycle.

At the scale of a local-scale environment (Figure 2.3(2)), studies lacking a closed system encounter a completely different set of challenges. Areas characterized by Se deficiency such as Finland or the U.K., for example, attempt amelioration by fertilising their crops with Se. While Finish authorities closely monitored resulting Se uptake into the food chain, little emphasis was placed on the environmental impact of such Se fertilization [Winkel et al., 2011]. Although dietary intake was increased by a factor of 4, crop uptake of Se was only 5 - 20 % of the annual Se application. It is still unclear what happened to the remaining 80 - 95 % of that applied Se, since it was thought to be immobilized, yet could not be accounted for when soil Se concentration was measured [Eurola et al., 2003]. This is disquieting, as Se is known to have a strong tendency to bioaccumulate – especially in aquatic biota [Winkel et al., 2011].

Modelling global-scale processes of the Se cycle (Figure 2.3(3)) faces another set of challenges entirely. While many lab-scale processes are understood in great detail, systematic measurement of environmental Se content and speciation throughout the world is sparse [Winkel et al., 2011]. But there have been attempts at creating a world Se map [Winkel et al., 2011] and Se flux estimates on a global scale have also been published [Haygarth, 1994, Amoroux, 2001].

To combine the benefits of lab-scale research with the applicability of local environmental-scale research, this study combines an interdisciplinary approach – botanical and geochemical – with a controllable, lab-scale biogeochemical model. This enables the simulation of multiple parameters of a local environmental-scale situation, while retaining detailed quantification of lab-scale precision.

2.6 Premises and hypotheses for this study

In reference to the classification of Figure 2.3, the first premise is that a lab-scale biogeochemical model with three compartments – plants, nutrient solution and soil minerals – will be an adequate compromise between a realistic reflection of environmental processes and practical labwork. Concerning the sudivision of these experiments, the second premise assumes that processes between each compartment can be studied separately between solution & plant and then soil & solution before synergistic effects are uncovered in combined experiments with all three compartments. This compartmentalization is what enhances the significance of results, since factors influencing Se transport between two compartments may be absent in the Se transport between two others. Furthermore, as explained earlier, smaller lab-scale compartments reduce the amount of influencing factors. The ability to quantify Se transfer beyond characterization of the Se transfer processes is crucial to creating a mass balance of Se in the Critical Zone. Therefore, the third premise is that results on Se transfer can be mathematically described, even if these mathematical descriptions might not quantify mono-causal processes.

Regarding the outcome of the experiments, the following hypotheses are included in the choice of experimental set-up, based on previous publications and knowledge of Se behaviour:

1. Se uptake of rice will be dependent on Se speciation and nutrient ion availability;
2. regarding the soil minerals, Se prevalence – and, therefore, also bioavailability – will be dependent on adsorption-desorption processes, which are influenced by Se behavior pertaining to mineral surface properties and soil solution properties;
3. in combined experiments, there will be synergistic effects not observable in single experiments.

3 Biochemistry – Se-uptake by *Oryza sativa*

The first part of this study is concerned with soluble Se and its uptake into and interactions with biota in the Critical Zone. While the biological components taking up Se could be either microorganisms, funghi, plants or animals such as insects or cattle – which all have a large impact on element cycles in the Critical Zone – this study focuses on biochemical properties as they pertain to the uptake of Se by plants. Given that in continental ecosystems Se is primarily introduced into the foodchain by plants, this approach targets the interface between abiotic Se and biotic Se compartments of the Se cycle.

3.1 Selenium in higher plants

Selenium content in plants generally reflects the levels of bioavailable Se in the parent soil, which is why most plant species contain less than 5 µg/g Se dry weight (DW) [Fernández-Martínez & Charlet, 2009]. Exceptionally high amounts of Se are taken up by Se-hyper-accumulators, like those of the genera *Astragalus* and *Stanleya*, which can accumulate up to 10 - 15 mg/g Se DW from soils containing only 2 - 10 µg/g Se DW [Virupaksha & Shrift, 1965]. It is believed that the majority of plants take up Se due to its similarity to sulfur (S), because, as of yet, Se has not been shown to be essential for the completion of the life cycle of plants [Läuchli, 1993, Pilon-Smits et al., 2009]. However, there have also been studies on possible allelopathic roles of Se-uptake by plants, such as protection from herbivores or pathogens [Quinn et al., 2008, Cappa et al., 2014].

Selenate is accumulated in plants against its electrochemical potential gradient. Research on this fact revealed that selenate is taken up actively across the root plasma mediated by a high-affinity sulfate transporter, which is regulated positively by O-acetylserine and negatively by sulfate and reduced glutathione [Asher et al., 1977, Arvy, 1993, Terry et al., 2000].

Sulfate and selenate uptake show both antagonistic and synergistic interactions when it comes to plant Se-uptake behavior. On the one hand, selenate competes with sulfate for uptake into plants and the presence of sulfate can inhibit Se-bioavailability. In Se-accumulators, on the other hand, selenate is taken up preferentially over sulfate in cases of high sulfate salinity [Terry et al., 2000]. Rice does not belong to the category of Se-accumulators. Nevertheless, rice plants have also been shown to take up Se preferentially in the presence of high sulfate [Bellet al., 1992, Terry et al., 2000].

Unlike selenate uptake, mechanisms of selenite uptake are still a matter of debate. In the past, studies suggesting passive uptake of selenite acknowledged having difficulties providing final evidence supporting their theory [Ulrich & Shrift, 1968, Asher et al., 1977]. Nonetheless, this theory is still being cited [Zhang et al., 2003]. However, more recent studies have reported uptake competition by phosphate as well as by sulphite [Hopper & Parker, 1999, Zhang et al., 2006, Li et al., 2007] and partial sensitivity to metabolic inhibitors [Li et al., 2007]. This suggests that active and specific mechanisms are responsible for selenite transport, with uptake likely to be governed by the phosphate transporter system [Li et al., 2007] and/or a silicon influx transporter system [Zhao et al., 2010].

Uptake of Se can also occur when Se is present in organic forms such as selenomethionine (SeMet) [Zayed et al., 1998]. There is also evidence that Se(0) has, until now, falsely been believed to be bio-unavailable and that Se(0)-nanoparticles might indeed be taken up as well [Fernández-Martínez & Charlet, 2009].

Translocation of Se within the plant strongly depends on the form of applied Se. Selenate was found to behave similarly to sulfate, entering plasma membranes of root cells unspecifically by the high-affinity sulfate transporters [Khan & Hell, 2014]. Throughout the plant, selenate then enters xylem transport vacuole membranes, inner plastides and tonoplasts unspecifically by further sulfate transporters [Khan & Hell, 2014] and is, therefore, quickly transported into shoots via xylem and phloem [Carey et al., 2012]. Its similarity to sulfate causes selenate to be reduced by the same pathways as sulfate as well, yielding selenide and finally selenocysteine (SeCys), which can be specifically methylated to produce the non-protein amino acid methyl-selenocysteine (MeSeCys) [LeDuc et al., 2004, Khan & Hell, 2014]. Selenite, on the other hand, is believed to be retained in the roots, but not as selenite [Läuchli, 1993]. Rather, selenite is transformed into organic compounds, such as SeMet, MeSeCys or SeCys, which are then distributed throughout the plant at a much lower rate, exclusively via phloem [Asher et al., 1977, Zayed et al., 1998, Terry et al., 2000, Sun et al., 2010].

3.2 Rice (*Oryza sativa*) as a model plant

The model plant of choice for our experiments is rice (*Oryza sativa*) of the cultivar *nihonmasari*. Rice, also known as the 'cereal crop of the world's poor' [Cantrell & Reeves, 2002], has been cultivated as a crop for more than 7,000 years. Currently, it sustains more than half of the world's population [Izawa & Shimamoto, 1996, FAOSTAT, 2012], most of whom are desperately poor [Cantrell & Reeves, 2002]. Though there are two cultivated rice species, *Oryza sativa* – also referred to as Asian rice – and *Oryza glaberrima* – referred to as African rice –, the predominant species *Oryza sativa* is the main staple of rice-growing countries as well as the main focus of subsequent research.

Oryza sativa is a diploid, annual, short-day monocotyledonous cereal plant which self-fertilizes [Izawa & Shimamoto, 1996]. What makes rice so interesting as a model plant is the fact that its genome is much shorter than comparable monocotyledonous cereals such as maize and wheat, although it shares much genetic similarity with these other crop-relevant plants [Cantrell & Reeves, 2002]. Due to its tremendous importance for human nutrition, the rice plant has been of great interest to research, leading to early genome sequencing [Goff et al., 2002] and detailed molecular mapping which, in turn, makes it an even better model plant for further research [Cantrell & Reeves, 2002].

However, the genetics of Se-accumulation are barely investigated, although it is well documented that inter- and intraspecific variations in Se accumulation in plants exists [Zhu et al., 2008]. Moreover, biofortification of food plants with Se is generally problematic. On the one hand, simultaneous fortification with Se together with other nutrient elements such as S is challenging due to uptake competition and metabolic crosstalk [Khan & Hell, 2014]. On the other hand, rice is able to take up Se beyond the recommended limits of human Se intake, so that fortification may lead to toxicity [Huang et al., 2013]. Therefore, our study was focused on clarifying antagonistic influences of nutrient addition and Se-fortification on one of human nutrition's most vital crops.

Previous research on Se-uptake into cereals has focused on agricultural water management [Li et al., 2010] and Se biofortification, such as uptake from Se-supplemented soils by barley [Gissel-Nielsen, 1973], wheat and ryegrass [Keskinen et al., 2010], durum wheat [Kikkert et al., 2013] and rice [Hu et al., 2002, Li et al., 2010]. In addition, mechanistic biomolecular studies on inner-plant transportation or single-ion competition have been conducted on excised barley roots [Leggett & Epstein, 1956, Ferrari & Renosto, 1972] and rice [Mikkelsen & Wan, 1990, Zhao et al., 2010, Carey et al., 2012].

Experimental set-ups vary greatly between these studies, depending on research focus. On the one hand, selenium addition to soils, which resemble the situation in real-world environmental conditions more closely, is a meaningful strategy to study biofortification. However, this approach is not appropriate to investigate the mechanisms of speciation or ion competition, because these are obscured by the complex interactions of Se and other nutrient components with multiple soil minerals. As major constituents of natural soils, organic matter, clay minerals and iron oxides or hydroxides, for example, add considerable sorption and complexation processes, which influence Se availability to the plant depending on pH-values and speciation of the added Se [Bar-Yosef & Meek, 1987, Balistrieri & Chao, 1990, Neal, 1995, Fernández-Martínez & Charlet, 2009]. Moreover, rice is unique in that it has specifically adapted to flooded conditions [Bellet al., 1992], where Se is most likely to be present as selenite [Zhao et al., 2010], elemental Se or selenide [Li et al., 2010]. Approaches such as these are intermediate- or long-term experiments (weeks or months), in which microbiological activity or plant growth-related changes create a dynamic equilibrium difficult to model in its complexity.

On the other hand, short-term excision experiments (hours or days) are the best solution for specifically focused bio-molecular studies on inner-plant transportion [Leggett & Epstein, 1956, Ferrari & Renosto, 1972]. However, plants in such experiments are usually exposed to Se after pre-cultivation in Se-free medium, which does not address the agronomically important aspect of germination and early seedling development in the presence of Se. Although the effect of Se on rice germination itself has been addressed [Liu & Gu, 2009], subsequent plant growth was not part of that study, so that an integrated view on early development in the presence of Se is lacking so far. A more holistic approach on the complexity of speciation-dependent uptake, translocation and distribution exists for wheat [Li et al., 2007]. Their focus, however, was on single-ion competition.

Therefore, neither Se-toxicity in early-stage rice seedlings nor nutrient supply effects thereupon have yet been explored. In this study, rice is considered to be most suitable for the plant-box studies, as it is an easily cultivated and well-known model organism with great relevance to human nutrition. This provides the basis for experiments on Se bioavailability and toxicity in early-stage seedling growth depending on Se concentration and Se species (Chapter 3.3.2), nutrient competition effects on Se uptake (Chapter 3.3.3) and influences on Se-bioavailability due to soil sorption processes dependent on mineral surface properties and Se species (Chapter 6.5c).

3.3 Experimental set-ups

Quantifiable characterization of an element in an intricate system such as the Critical Zone requires a maximum of controllable parameters concerning in- and output to and from this system.

To determine the variables influencing Se uptake into plants, it was essential to create an environment with ideal growth conditions so that ailments due to external factors unrelated to Se-uptake, such as nutrient deficiency, lack of light or insufficient temperature and water provisions could be excluded. Otherwise, signs of malnutrition in plants might be misinterpreted as Se toxicity symptoms or malnutrition might induce other metabolic pathways in the plants uncommon in healthy agricultural crops.

Plant requirements provided in this study are: sufficient photosynthetically usable light, CO_2 partial pressure, optimal temperature and humidity appropriate for the species' growth, all of which can be adjusted in a climate chamber or restricted area green houses with special lighting.

Furthermore, plants require water, nutrients and a substrate firm enough to offer rooting support for the plant. This has direct implications for this study. To determine plant-dependent parameters in the Se cycle in the Critical Zone, four plant experiments were conducted differing in nutrient supply and germination Se exposure (Exp. A1, A2, A3 and A4), the first of which is an open culture, while the following three are closed plant-box cultures.

3.3.1 Exp. A1 – open rice cultures in Se-nutrient solution

In this first plant-based experiment, emphasis was placed on the nutritional and water supply requirements of *Oryza sativa* plants. These were monitored and quantified for future closed-system-experiments.

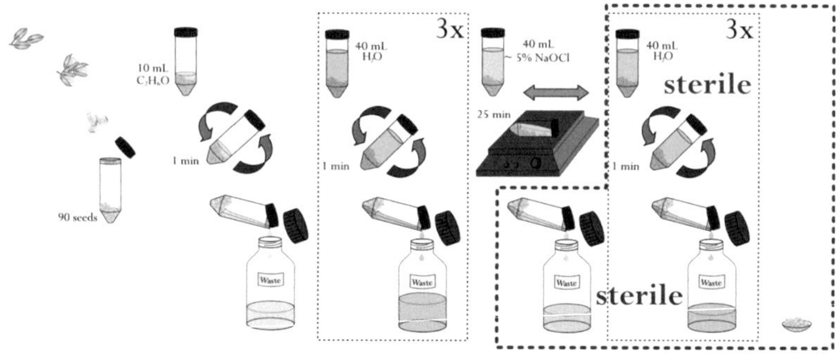

Figure 3.1: Sterilization procedure for 40 - 100 rice caryopses in a 50-mL PP-centrifuge tube (VWR 525-0155): removal of the husks (palm husker, Ambala Associates), pre-sterilising in ethanol for 1 min before decanting (10 mL of 70 % ethanol, mixed as 700 mL p.a. ethanol absolute (Roth, Art. No. 9065.4) and 300 mL of double-deionized water), rinsing with double-deionized water three times, sterilising in 40 mL of 5 % NaOCl (as 400 mL NaOCl stock solution (Roth Art. No. 9062.3, 12 % active Cl) and 600 mL double-deionized water), shaking at 120 rpm for 25 min; subsequent NaOCl decanting and 3 times rinsing in 40 mL of autoclaved double-deionized water are done under sterile conditions

To this purpose, cultures of *Oryza sativa* of the cultivar of *nihonmasari* were sterilized, grown on agar to a size large enough for re-potting, and then transferred to the actual experimental set-up, which consisted of flower pots filled with chemically nearly inert quartz sand. This experiment focused on plant uptake of nutrients and water provided by a specially devised nutrient solution.

Rice pre-cultivation in plant-boxes

The sterilization of caryopses for all experiments in this study followed the same standardized explant-surface sterilization protocol used in most laboratories [Oyebanji et al., 2009], which was adjusted empirically in terms of durations and solution amounts as shown schematically in Figure 3.1.

(a) (b) (c)

Figure 3.2: Seedlings grown in agar in plant-boxes;
a: experimental set-up schematic (Inkscape);
b: Photo taken 30.04.2012 (3 days after sowing);
c: Photo taken 07.05.2012 (10 days after sowing)

In preparation for planting, two assembled plant-box sets, each with a lower half (Sigma Aldrich, Art. No. V8380) and an upper half (Sigma Aldrich, Art. No. V8505) connected with a coupler (Sigma Aldrich, Art. No. C0667), were autoclaved in a Systec VE-95 at 123.5 °C and 210.4 kPa for a sterilization time of 15 min along with 200 mL 0.4 %

(= 0.8 g) phytoagar (Duchefa Direct, CAS # 9002-18-0) in a 500 mL Schott duran glass bottle. Under cleanbench conditions, 100 mL of the autoclaved, warm liquid agar were poured into the smaller half of each of the magenta box sets and left to cool down to room temperature. Twenty caryopses were planted into each box in rows of four with the embryo facing upwards just below the surface of the cooled agar.

Boxes were sealed under cleanbench conditions and then kept in a PAR (**P**hotosynthetically **A**ctive **R**adiation)-light chamber for 2 weeks to promote optimal growth (Figure 3.2).

Transfer into open cultivation in quartz sand

As shown in Figure 3.2 (right), the first plants reached plant-box height (ca. 13 cm) after 10 days. Therefore, it was not deemed advisable to keep them growing in such boxes for longer than a maximum of 2 - 3 weeks. Furthermore, with only phytoagar as a substrate, nutrient supply was not sufficient for further growth without risking malnutrition.

For transfer into a new experimental set-up, the following aspects were taken into account:

- available root space, which was varied by using two different pot sizes: ca. 300 mL and 900 mL flower pots (geli Thermo Plastic, PP Pflanzkübel Blumentopf Standard rund)
- inert substrate choice; to quantify nutrient and water uptake, quartz was chosen as a geochemically fairly inert substrate
- nutrient supply; to be able to control details of the nutrient solution composition, a specifically designed nutrient solution was required
- *Oryza sativa*-specific climate was attained by growing plants under greenhouse conditions

Out of the 40 seedlings planted into agar on April 27th, 2012, 35 developed normally and were re-potted into quartz sand (Dorsillit No. 98 0.1 - 0.6 mm) on May 11th, 2012.

(a) (b) (c)

Figure 3.3: Preparation of open culture experiments
a: flower pot taped with gauze;
b: watering the freshly re-potted seedlings with 35 mL of nutrient solution;
c: reference plants with nutrient solution filled to 400 mL

To keep the quartz sand from flowing out through the water holes of the plant pots due to liquefaction, these holes were taped with fishing gauze (Figure 3.3a).

The quartz sand was moistened with once-deionized water before filling ca. 750 mL into the large pots and ca. 250 mL into the small pots. As shown in Figure 3.3b, each of the 35 seedlings was planted separately into a pot and then watered with 35 mL of nutrient solution (preparation described below). After that, pots were organized into six green drainage boxes and each plant was labelled with a box number and plant letter in alphabetical order (Figure 3.4).

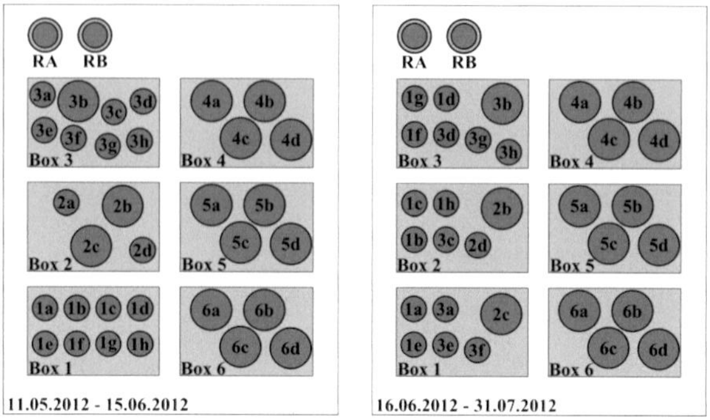

Figure 3.4: Schematic drawing of plant order and labelling in Exp. A1 (RA, RB = reference plants A and B);
left: Plant order 11.05.2012 - 15.06.2012;
right: change in plant order after change in watering regime 16.06.2012 - 31.07.2012

Nutrient supply and watering regime

The nutrient solution was composed specifically for this experiment, with nutrient content and concentrations based roughly on the Murashige & Skoog medium [Murashige & Skoog, 1962] and nutrient solutions used in other, similar experiments [Bellet al., 1992] [Clement et al., 1977] [Asher et al., 1977]. Although rice can take up Fe(III) by using special chelating agents, Fe(II), is more soluble and more bioavailable [Ogo et al., 2007]. Therefore, Fe is the only element, which was added as the organic complex ferric citrate; all other elements were added as inorganic compounds, primarily as sulfates, nitrates or phosphates.

Table 3.1 lists the compounds of this nutrient solution, which were all dissolved in double-deionized water. $CuSO_4$, $ZnSO_4 \cdot H_2O$ and $MnSO_4 \cdot H_2O$ were dissolved together in 10 mL of double-deionized water. Except for ferric citrate, which was heated in an oven to 80 °C, all components were dissolved at room temperature.

Table 3.1: Nutrient solution compounds of Exp. A1; inweight for target concentration of 100 L and required minimal dissolution volume of double-deionized water for the concentrate

compound	product name	final solution concentration [mg/L]	in-weight for 100 L [mg/L]	disolving volume [mL]
$Ca(NO_3)_2 \cdot 4H_2O$	Sigma-Aldrich, p.a. C1396-500G	236.145	23.596	40
K_2SO_4	Merck, p.a. 1.05153.0500	65.346	6.532	120
$MgSO_4 \cdot 6H_2O$	Fluka, p.a. 00627	74.248	7.435	30
KH_2PO_4	Merck, p.a. 1.04873.1000	13.608	1.360	40
H_3BO_3	Sigma-Aldrich, p.a. B6768-500G	0.495	0.049	25
$CuSO_4$	Sigma-Aldrich, p.a. 61230-100G	0.032	0.003	10
$ZnSO_4 \cdot H_2O$	Sigma-Aldrich, p.a. 96495-250G	0.090	0.001	10
$MnSO_4 \cdot H_2O$	Sigma-Aldrich, p.a. M7634-100G	0.034	0.003	10
$CaCl_2$	Merck, p.a. 1.02378.0500	5.549	0.555	20
$Na_2MoO_4 \cdot 2H_2O$	Merck, p.a. 1.06521.0100	0.012	0.002	15
$C_6H_5O_7Fe$	Sigma-Aldrich, p.a. F6129-250G	12.247	1.224	200

The nutrient solution was stored at -20 °C as two concentrated stock solutions (factor 200), separating Ca-containing components and sulfates to avoid $CaSO_4$ precipitation.

To be able to quantify the amount of water required by the plants through evapo-transpiration, two plants were chosen as reference plants, RA and RB. As shown in Figure 3.3c, these plants in small pots were hung into 500 mL glass beakers. Each time the plants were watered, the amount of nutrient solution needed to bring the solution level back to the 400 mL line was noted for both plants.

Unlike the large drainage boxes, the small pots fit into the beaker without leaving the solution surface openly exposed to evaporation, therefore making it possible to determine the amount needed solely due to evapo-transpiration directly related to the plant. These amounts were used to determine nutrient solution volume for all plants.

Using a Schott bottle cap (27 mL ±3 mL), plants were watered three times a week (Mondays, Wednesdays and Fridays) with nutrient solution, which was both the plant's sole source of water and only source of nutrients. Half of the plants were watered with 27 mL/day of nutrient solution, while the other half was watered with 55 mL/day of nutrient solution to quantify vitality differences due to lower nutrient uptake. Plants were grown in this fashion from May 11th, 2012 to July 15th, 2012.

Addition of Se

The open culture experiment had to be modified due to some plants showing fungal infection, loss of vitality and overall inhomogenous plant development (Figure 3.5). Plants were categorized into "low nutrient solution supply" and "high nutrient solution supply" as well as into "big pot" and "small pot".

(a) (b)

Figure 3.5: Photos of A1 experimental set-up for Se addition to open culture rice plants
a: beginning of experiment (Photo taken 14.05.2012)
b: plant growth until addition of Se (Photo taken 27.07.2012)

On July 16th, 2012, 65 days after transfer into open cultures, plant pots were re-arranged (Figure 3.4) to distribute plants of each category evenly among the drainage boxes. A 1000 mg/L Se-stock solution was prepared of sodium selenate decahydrate, $Na_2SeO_4 \cdot 10H_2O$ (VWR BDH

Prolabo, 302113L) in 100 mL of double-deionized water. As shown in Figure 3.5, for the following 2 weeks, boxes 3 and 4 and reference plant RB were watered with nutrient solution, boxes 2 and 5 were watered with nutrient solution spiked with 2.5 mg/L Se and boxes 1 and 6 as well as reference plant RA were watered with nutrient solution spiked with 5 mg/L Se. Plants were harvested on July 31st 2012 as described below (Chapter 3.4.1).

3.3.2 Exp. A2 – plant-box rice cultures in Se-agar

The second set of experiments focused on Se uptake directly during germination and toxicity in early-stage growth of rice plants. Since quantifying input and output is easiest in closed systems, a plant-box solution was chosen. As mentioned in Exp. A1 (Chapter 3.3.1), this restricted the plant growth period to a maximum of three weeks.

For the first set of plant-box experiments (Figure 3.6a), 10 assembled Magenta box sets (Sigma Aldrich, Art. No. V8380, V8505 & C0667) were autoclaved in a Systec VE-95 at 123.5 °C and 210.4 kPa for a sterilization time of 15 min, together with 1200 mL 0.4 % (= 4.8 g) phytoagar (Duchefa Direct, CAS # 9002-18-0), a 100-mL measuring cylinder and 200 mL of double-deionized water. Ninety rice caryopses were sterilized according to the procedure described above in Chapter 3.3.1.

From a 1000 mg/L Se stock solution (in double-deionized water, either $Na_2SeO_4 \cdot 10H_2O$ [VWR BDH Prolabo 302113L], or Na_2SeO_3 [AlfaAesar 012585], respectively), 10 mL of a 100 mg/L Se solution in double-deionized water was sterilized in the cleanbench using a 0.22-µm sterilization filter (Omnifix T550.1, Rotilabo P 666.1). Under cleanbench conditions, plant-boxes were opened and 100 mL of the autoclaved liquid 0.4 % agar (ca. 50 °C) was poured into the smaller half of each of the plant-box sets.

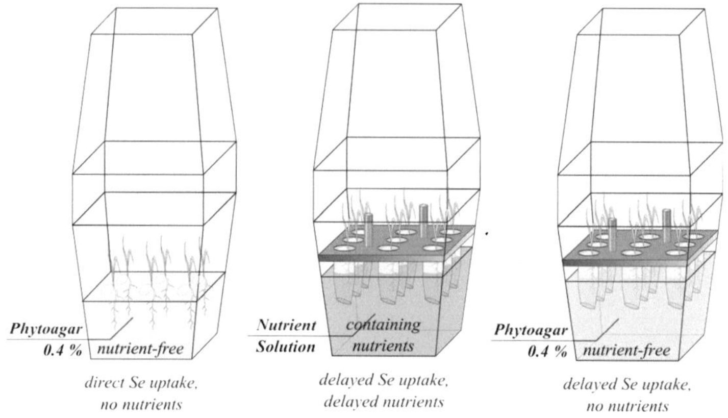

(a) (b) (c)

Figure 3.6: Schematics of Exp. A2, A3 & A4
a: Se-agar experiments contained 6 rice plants grown in 100 mL Se-agar for 16 days with Se concentrations of 0, 5, 10, 25, 50, 100, 250, 500, 1000 and 2500 µg/L Se added as either selenate or selenite.
b: Se-nutrient solution experiments contained 9 rice plants grown for 19 days, sprouting first in Se-free agar in Eppendorf tubes before reaching 170 mL of Se-spiked nutrient solution (0 - 2500 µg/L Se) after ca. 5 days.
c: Se-agar experiments with delayed Se-uptake contained 9 rice plants grown for 19 days, sprouting first in Se-free agar in Eppendorf tubes before reaching 170 mL Se-spiked agar (0 - 2500 µg/L Se) after ca. 5 days.
All three sets of experiments were kept in a climate chamber with an 8h-short-day rhythm, 1 h of each dusk and dawn and a temperature of 28 °C during the day period and 22 °C during the night period.

While the agar was still liquid, though not hotter than 50 °C, the sterilized Se-solution was added to yield the appropriate final concentrations (0, 5, 10, 25, 50, 100, 250, 500, 1000 and 2500 µg/L Se). Using a 5-mL pipette tip, Se was homogenized in the agar before sampling

2 mL from each box into the allotted 20-mL patho-vessels (Böttger 08-313-1001), which had been prepared with 1 mL of nitric acid (VWR 20429.320 p.a. subboiled to suprapure quality) and 1 mL of double-deionized water and weighed (including the lid and tape covering the marker-written labels to avoid washing off by sterilising ethanol). These were weighed to determine the exact weight per volume of the agar and kept at 4 °C. Plant-boxes were then left to cool until the agar was firm. As described in Chapter 3.3.1, 6 rice plants were sown into the agar and the boxes were sealed.

For the duration of 16 days, plant-box cultures were kept in a climate chamber (YORK Refrigeration) at 70 % humidity, and a sub-tropical 8h-short-day rhythm, 1 h of dusk and dawn each and a temperature of 28 °C during the day period and 22 °C during the night period. This time frame was calculated as 2 days of sprouting and 14 days of Se-uptake. Plant growth was monitored by measuring plant height on days 5, 7, 9 and 12 for each plant.

After 16 days, magenta box cultures were harvested as described below (Chapter 3.4.1). For each experimental set-up, three independent biological replicas were conducted, one of which was conducted by bachelor student Matthias von Brasch as part of his thesis under my supervision.

3.3.3 Exp. A3 – plant-box rice cultures in Se-nutrient solution

In the second plant-box experiment (Figure 3.6b), plants were grown in the presence of all essential nutrients, similar to the nutrient solution used in Exp. A1 (Chapter 3.3.1). Because these closed plant-box systems were not to be opened during the entire duration of the experiment, nutrient solution composition had to be modified to that of Exp. A1

(Chapter 3.3.1) to provide for the growing plants for the entire duration of the experiment.

Optimal nutrient solution composition (Table 3.2) was obtained empirically in preliminary experiments. The nutrient solution was stored at -20 °C as two stocks (factor 200), separating Ca-containing components and sulfates to avoid $CaSO_4$ precipitation.

Table 3.2: Optimized nutrient solution compounds of Exp. A3, inweight for target concentration of 100 L and required dissolution volume of double-deionized water

compound	product name	final solution concentration [mg/L]	in-weight for 100 L [mg]	disolving volume [mL]
$Ca(NO_3)_2 \cdot 4H_2O$	Sigma-Aldrich, p.a. C1396-500G	1180.745	118.0745	200
K_2SO_4	Merck, p.a. 1.05153.0500	65.346	6.532	120
$MgSO_4 \cdot 6H_2O$	Fluka, p.a. 00627	74.248	7.435	30
KH_2PO_4	Merck, p.a. 1.04873.1000	54.432	5.443	160
H_3BO_3	Sigma-Aldrich, p.a. B6768-500G	0.495	0.049	25
$CuSO_4$	Sigma-Aldrich, p.a. 61230-100G	0.064	0.006	10
$ZnSO_4 \cdot H_2O$	Sigma-Aldrich, p.a. 96495-250G	0.135	0.014	10
$MnSO_4 \cdot H_2O$	Sigma-Aldrich, p.a. M7634-100G	0.204	0.020	20
$CaCl_2$	Merck, p.a. 1.02378.0500	5.549	0.555	20
$Na_2MoO_4 \cdot 2H_2O$	Merck, p.a. 1.06521.0100	0.018	0.002	25
$C_6H_5O_7Fe$	Sigma-Aldrich, p.a. F6129-250G	12.247	1.224	200

Ten assembled Magenta plant-box sets (Sigma Aldrich, 2 x V8505 & 1 x C0667) were prepared (Figure 3.6b) with a specially designed Eppendorf tube tray holder for 9 x 1.5-mL tubes (1.5-mL Eppendorf Safe-Lock Microcentrifuge Tube, 0030120086). The plant-box sets, 90 1.5-mL Eppendorf tubes, 250 mL 0.7 % phytoagar, 2.5 L nutrient solution (preparation as given above), a 100-mL measuring cylinder and 250 mL double-deionized water were autoclaved. Sterile rice caryopses as well as $Na_2SeO_4 \cdot 10H_2O$ and Na_2SeO_3-standards were prepared as described above (Chapter 3.3.2). After autoclaving, 1.5-mL reaction tubes (Eppendorf, Hamburg) were placed into a tray-holder under the cleanbench and filled to the brim with the still-liquid 0.7 % phytoagar. Only after the agar had solidified, were the lid and 0.5 cm from the bottom of each reaction tube cut off using sterilized scissors. Each tube was then loaded with one sterilized caryopse.

Plant-boxes were disassembled under the cleanbench, Eppendorf trays removed and 200 mL of nutrient solution filled into each lower half of the boxes. Specified concentrations of Se (0 - 2500 µg/L Se, as described in Chapter 3.3.2) were obtained by diluting the stock into the nutrient solution. After mixing the solutions with a pipette tip, 2 x 15 mL of each solution were sampled for subsequent analysis into 20-mL patho-vessels (Böttger 08-313-1001) - 15 mL were frozen for later IC measurement (Chapter 3.4.4), the other 15 mL were acidified with 50 µL of HNO_3 (VWR 20429.320 p.a. sub-boiled) and stored at 4 °C for later ICP-MS measurement (Chapter 3.4.4). For plant-box assembly, each Eppendorf tray was loaded with 9 rice-bearing Eppendorf tubes and placed into the plant-box with the lower half of the tubes submerged in nutrient solution before closing the plant-box. These were kept under the same climate chamber conditions as described above (Chapter 3.3.2), keeping the boxes closed until harvest (Chapter 3.4.1). Each experimental set-up was repeated independently three times.

3.3.4 Exp. A4 – plant-box rice cultures in Se-agar with delayed Se-uptake

For the purpose of comparison, an experiment was devised to separate the two parameter differences of both previous plant-box experiments (Figure 3.6c). The non-nutrient-bearing agar of experiment A2 (Chapter 3.3.2) was combined with the uptake delay of the nutrient solution plant-boxes from experiment A3 (Chapter 3.3.3).

Similarly to the nutrient solution experiments, plant-boxes were prepared with trays of 9 x 1.5-mL Eppendorf tubes filled with 0.7 % phyto-agar, each planted with one rice plant. Instead of nutrient solution, tray holders were placed into 170 mL liquid 0.4 % agar spiked with 0 - 2500 µg/L Se (as described above Chapter 3.3.3) with the lower half of the tubes submerged in agar before closing the plant-box. Boxes were left until the agar solidified and then placed in the climate chamber described above (Chapter 3.3.2). Plants were harvested after 19 days as described below (Chapter 3.4.1). This experiment was conducted only once for selenate and selenite, each.

3.4 Sample preparation and analytical methods

Because harvest, digestions and subsequent HG-FIAS analysis were alike for all plant experiments, as well as the IC- and ICP-MS analysis of all nutrient solution samples, the following chapter will detail these methods once.

3.4.1 Plant harvest

After photo-documentation using a Samsung S4 mini cell phone camera, shoot height and 2nd leaf length were measured; then harvested plant material was cleaned under running mono-deionized water to remove quartz sand substrate (Exp. A1, Chapter 3.3.1), agar (Exp. A2, A4 Chapter 3.3.2, 3.3.3) or nutrient solution residue (Exp. A3, Chapter 3.3.3). Plants were then rinsed with double-deionized water before being air-dried for 2 h at standard lab conditions (22 °C, 1 atm, 30 % humidity) and separated into shoot and root above the caryopse (Figure 3.7a) using a pair of clean scissors. Each sample was weighed (Sartorius basic BA 1105) to determine fresh root and shoot weight.

Using a Christ Alpha 1-4, 100400 freeze-dryer, plant material was freeze-dried at 0.05 mbar and -20 °C for 24 h. After warming up to lab conditions, samples were weighed anew to determine plant water loss. Samples of the open culture experiments were milled using an agate disc mill (Scheibenschwingmühle-TS, Siebtechnik) to pulverize and homogenize the entire material of each sample. In the case of plant-box studies, root and shoot plant-box bulk samples were cut into 0.2 cm pieces using a clean pair of scissors. The use of Teflon scissors and Teflon forceps was discontinued due to massive static charge.

All plant samples were stored in 20 mL patho vessels (Böttger 08-313-1001) at lab conditions until digestion (Chapter 3.4.2). To sample the viscous agar during the preparation of Exp. A2 (Chapter 3.3.2), the top 1 cm of a 5-mL pipette tip was cut off and using an adjustable 0.5 - 5-mL pipette, amounts of 2 mL of the still-liquid agar were directly pipetted and weighed (Sartorius basic BA 1105) into 20-mL patho-vessels (Böttger 08-313-1001) prepared with an in-weighed 1 mL of nitric acid (VWR 20429.320 p.a. subboiled to suprapure quality) and 1 mL of double-deionized water. To sample the cooled agar at the end of the experiment, agar was sampled directly from the plant-box after stirring it with a pipette tip. Due to the large amount of air within the cooled, stirred agar, patho-vessels were placed on a balance (Sartorius 1712MP8) and a weight of 2 g was sampled, rather than a volume of 2 mL. All plant and agar samples were stored in the refrigerator at 4 °C until digestion.

From the nutrient solution experiments, 2 x 15 mL of nutrient solution was sampled from the plant boxes into 20-mL patho-vessels (Böttger 08-313-1001) - 15 mL were frozen for later IC measurement (Chapter 3.4.4), the other 15 mL were acidified with 50 µL of HNO_3 (VWR 20429.320 p.a. sub-boiled) and stored at 4 °C for later ICP-MS measurement (Chapter 3.4.4).

3.4.2 Microwave digestions

For the plant digestion, small adjustments were made to the procedure described in previous studies [Bellet al., 1992]. In-weights (Sartorius 1712MP8) of 0.1 g or less were filled into Teflon vessels (Figure 3.7b,c,d) of an MLS START 1500 Microwave System preconditioned with 3 mL of double-deionized water and 3 mL HNO_3 p.a. (VWR 20429.320) at 150 °C for 1 h on a hotplate (Ceran 500, LHG).

Samples were digested with 1 mL of double-deionized water, 3 mL HNO_3 conc. (VWR 20429.320 p.a. sub-boiled) and 1 mL of 30 % H_2O_2 (p.a. Rotipuran 8070.3). Each batch of 10 digestion samples included one Teflon vessel blank and one plant standard (0.1 g NBS SRM 1567a Wheat Flour) to verify digestion quality.

(a) (b) (c) (d)

Figure 3.7: Plant digestion with a microwave system;
a: separation of shoot and root (Inkscape illustration);
b: Teflon lid fixation onto Teflon vessels with pressure ring and vessel holders;
c: Teflon vessels in holders ready to be mounted;
d: Teflon vessels mounted in microwave

Applying the temperature program shown in Table 3.3, samples were digested in the microwave at up to 220 °C and 1000 W for 33 min. To avoid volatile Se loss, samples were left in the microwave overnight to cool completely.

All digested samples, standards and blanks were quantitatively transferred into a 10-mL glass flask with a glass funnel and filled to volume with double-deionized water. Samples were then transferred and stored in 15-mL-centrifuge tubes (VWR 525-0149) in the refrigerator at 4 °C until analysis.

Agar digestions were prepared similarly to plant digestions, albeit with the additional difficulty of transferring the agar sample quantitatively into the Teflon vessels. This was achieved by using the remaining digestion ingredients (2 x 1 mL HNO_3 conc. and 1 mL of 30 % H_2O_2) to rinse all of the agar sample into the Teflon vessels.

Table 3.3: Digestion temperature program for microwave system START 1500

	t(min)	**T (°C)**	**E (W)**
	3	75	600
	8	130	700
	10	210	1000
	12	220	1000
	45	cooling	
total	1 h 18 min		

3.4.3 Plant-Se and agar-Se analysis with HG-FIAS

Se-content of digestion samples was analysed with HG-FIAS (Hydride Generation Flow Injection Atomic Absorption Spectroscopy) with a Perkin Elmer AAnalyst200 coupled with a FIMS-400 Hydride Generation System and Autosampler. Before measurement, a reduction of selenate to selenite was necessary for the hydride to be able to form H_2Se, which could then be atomized in the quartz tube and measured [Bye & Lund, 1988]. To achieve concentrations within calibration range, samples were diluted by a factor of 2 to 1200 in 6 mol/L HCl (Merck, 37 %, low-Hg HCl 1.13386.2500). From a 1000 mg/L Se standard solution (Roth Rotistar ICP), 10 mL of calibration standard (1000 µg/L Se) were prepared in a 15-mL centrifuge tube (VWR 525-0149). A sample volume of 3 mL each in 15-mL centrifuge tubes (VWR 525-0149) and the calibration standard were reduced to selenite for 15 min in a water bath (Dinkelberg analytics, E30U) pre-heated to 75 °C.

After cooling, calibration concentrations of 0.5, 0.75, 1, 2, 4, 5, and 6 µg/L Se (for open culture samples: 0, 0.5, 1, 2 and 5 µg/L) were prepared with 1 mol/L low-Hg HCl from the reduced Se solution. Then all samples were diluted to 12 mL with double-deionized water before measurement. Using a reducing agent of 0.2 % $NaBH_4$ (Merck 8.06373.0500) in 0.05 % NaOH solution, a carrier solution of 1 mol/L HCl (Merck, 37 %, low-Hg HCl 1.13386.2500) and an argon carrier gas flow rate of 50 - 70 mL/min, Se concentration was measured with 5 replicates (3 for open culture samples) as the peak height of atomic absorption at a wavelength of 196.2 nm (background subtraction: off, a 260-V EDL-lamp, slit width: 2.7/2.3, quartz cell temperature: 900 °C).

To verify measurement quality, a multi-element drinking water standard (PromoChem Trace Metals QCP 050-1 and QCP 050-2 combined, with 252 µg/L Se) was included with a dilution factor of 120, as well as a reduction blank of 6 mol/L HCl. For the purpose of quality control, within each measurement, all standards were re-measured as samples after every 10 samples for linear interpolation of drifts and the trace metal standard was measured at the beginning and end of the measurement.

3.4.4 Nutrient solution analysis with IC and ICP-MS

Ion chromatography (IC) analysis of anions was conducted using a Dionex ICS 1000 with an IonPac AS14 column coupled with an IonPac AG14 pre-column. The eluent consisted of a mixture of 3.5 mmol/L Na_2CO_3 and 1.0 mmol/L $NaHCO_3$ with an eluent flow rate of 1.10 mL/min. Frozen nutrient solution samples were thawed prior to measurement and diluted by a factor of 4 with double-deionized water before being filled into IC-vials covered with filter caps.

A multi-ion IC-standard calibration solution (Alfa Aesar, Specpure) containing fluoride, chloride, bromide, nitrate, phosphate and sulfate in proportions of 1 : 2 : 4 : 4 : 6 : 4, respectively, was diluted to calibration standards containing 1, 2.5, 5, 10 and 20 mg/L fluoride. Using an Anion Self-Regenerating Suppressor (ASRS 300), conductivity of samples was detected at an applied current of 25 mA for an injection volume of 25 µL.

Computer-aided peak analysis over the detection span of 14 min was manually adjusted for optimal standard comparability. Sample standard deviation (SD) was calculated by applying the 1σ percentile of repeated calibration solution measurements to all samples for each ion. A river water standard (BATTLE-02, Environment Canada; 42.4 mg/L Cl^-, 0.194 mg/L F^-, 149 mg/L SO_4^{2-}) was measured with a dilution factor of 2 to confirm analytical accuracy. For the purpose of quality control, within each measurement batch, one calibration standard was re-measured as a sample after every 8 - 12 samples for linear interpolation of drifts and the river water standard (to determine analytical reproducibility) was measured once per measurement batch.

Analysis of ^{11}B, ^{23}Na, ^{25}Mg, ^{26}Mg, ^{31}P, ^{39}K, ^{42}Ca, ^{43}Ca, ^{55}Mn, ^{56}Fe, ^{63}Cu, ^{65}Cu, ^{64}Zn, ^{66}Zn, ^{25}Mg, ^{77}Se, ^{78}Se, ^{82}Se, ^{95}Mo, ^{98}Mo, was performed using an inductively coupled plasma mass spectrometer (ICP-MS) X-Series 2 (Thermo Fisher Scientific). Five mL of sample were diluted by factors of 2 - 10 in 1 % subboiled HNO_3. Each sample was spiked with 50 µL of internal standard (10 µg/L Sc, Merck 1.70349.100; 10 µg/L Rh, Merck 1.70345.0100; 10 µg/L In, Merck 1.70324.0100; 10 µg/L Tm, Merck 1.70361.0100), which were measured as ^{45}Sc, ^{103}Rh, ^{115}In and ^{169}Tm. Calibration was carried out using the ICP multi-element standard solution VI Certipur (Merck 1.09493.0100) in dilution factors of 20,000, 10,000, 4000, 2000, 1000, 400, 200, 100, 20 and 10. Added to the multi-element standard, P (Merck 1.70340.0100) was calibrated with 5, 10, 25, 50, 100, 250, 500, 1000, 5000 and 10,000 µg/L. Calibration was extended to 5 and 10 mg/L Na (Merck 1.70353.0100), Mg (Merck 1.703331.0100), K (Merck 1.70342.0100) and Fe (Merck 1.70326.0100), as well as 25 and 50 mg/L Ca (Merck 1.70308.0100) using single element standards. Verification of analysis quality was achieved by measuring a trace metal standard CRM-TMDW-A (High Purity Standards) with a dilution factor of 5. For the purpose of quality control, within each measurement batch, one standard was re-measured as a sample after every 8 - 12 samples for linear interpolation of drifts and the trace metal standard was measured at the beginning and end of the measurement batch.

Gas flow of the argon sample carrier gas (nebulizer gas) was 0.87 L/min, flow of argon fuel gas (auxiliary gas) was 0.68 L/min and flow of argon cooling gas was 13 L/min. Each sample was measured in three different modes. First, a survey scan of 10 sweeps with a dwelling time of 13 seconds was performed to obtain a general overview of the sample's content. Then, the main peak jump was determined by measuring in CCT-ED mode (Collision Cell Technology – Energy Discrimination), with 3 times 40 sweeps and a dwelling time of 14 seconds for the remaining elements Na, P, K, Mn, Fe, Cu, Zn and Se. In CCT-ED measuring mode, the detection limit was lowered by including a collision cell between the plasma torch and the quadrupole mass spectrometer, in which collision gas, a mixture of 92 % He and 8 % H_2, was injected at an intercepting angle, colliding with the ionized sample gas. Compound clusters between argon and hydrogen with themselves, each other and sample elements were thereby slowed or destroyed and did not interfere with the actual sample measurement. After that, the main peak jump was measured for the elements B, Mg, Ca and Mo in standard measuring mode with 3 times 40 sweeps of 25 seconds dwelling time per isotope and 30 seconds of acquisition per run. In solutions, in which only Se was measured, 3 times 40 sweeps with a dwelling time of 25 seconds per isotope resulted in an acquisition time of 14 seconds per run.

3.5 Results for the Se treatment of rice plants

In the following chapter, results on the Se content of rice shoots and roots are given for each of the experiments listed above and combined with results on further parameters of plant health, such as plant mass, shoot height and heights of the second leaf.

3.5.1 Exp. A1 – open rice cultures in Se-nutrient solution

Quality measures for experiment A1 (Chapter 3.3.1) showed that harvested plant yield was 91 % of transferred seedlings. Digestion standard retrieval was 85 % (±24 %) and drinking water standard retrieval during HG-FIAS measurement was 105 % (±9 %).

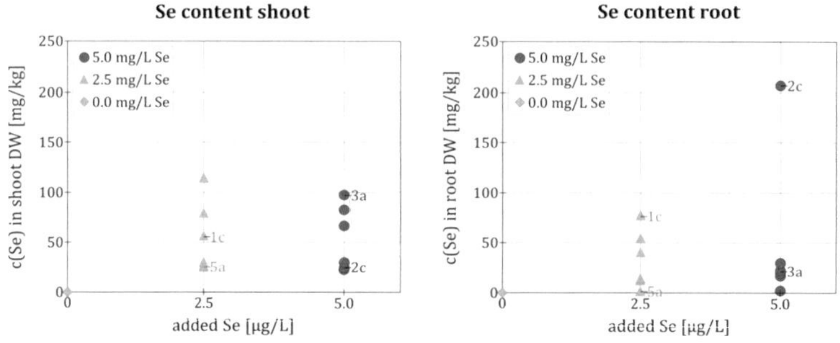

(a) (b)

Figure 3.8: Plant Se content after addition of 0, 2.5 or 5 mg/L Se to open culture rice plants for 2 weeks after 63 days of Se-free growth
a: Se content in shoots
b: Se content in roots

Selenium concentrations for the additions of 2.5 or 5 mg/L measured in the rice plants of experiment A1 ranged from 23 to 113 mg/kg DW for shoots (Figure 3.8a) and 4 to 207 mg/kg DW for roots (Figure 3.8b). Mean values and standard deviation (±1s) did not differ significantly between shoot Se-uptake after 2.5 mg/L Se addition (55 mg/kg DW, ±35) and shoot Se-uptake after 5 mg/L Se addition (51 mg/kg DW, ±31). Similarly, no significant difference was found in the mean values between root Se-uptake after 2.5 mg/L Se addition (40 mg/kg DW, ±27) or after 5 mg/L Se addition (48 mg/kg DW, ±71). Standard deviation was higher for the values of 5 mg/L root uptake, due to the outlier 2c, which had a very high Se content compared to all other samples (Figure 3.8b). When excluding the outlier 2c, however, mean root uptake was lower for 5 mg/L Se addition (21 mg/kg DW, ±10) compared to 2.5 mg/L Se addition. Therefore, differences in Se shoot content were negligible for both Se concentrations, but not for root Se content.

Here, addition of 5 mg/L Se resulted in only 53 % of root content compared to the addition of 2.5 mg/L. Furthermore, Se content of the root was lower than in the shoot. Root-Se for the addition of 2.5 mg/L Se was 72 % of shoot-Se and root-Se for the addition of 5 mg/L is 42 % of that in the shoots, excluding the outlier 2c.

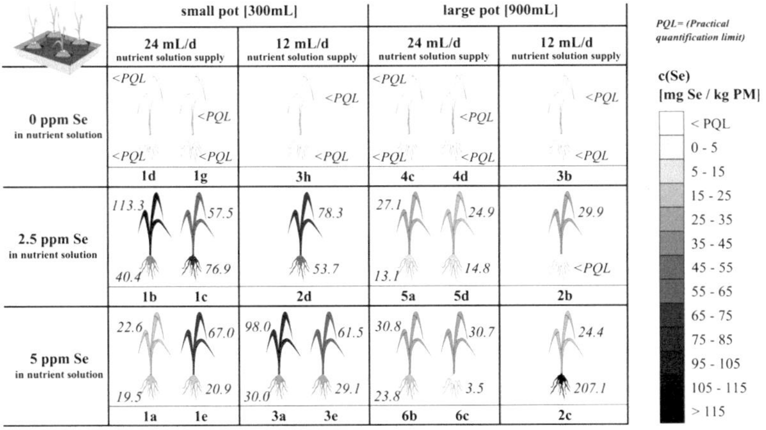

Figure 3.9: Single-plant Se content in shoots and roots of experiment A1

When plants were differentiated according to experimental conditions (Figure 3.9), a Se-content distribution pattern emerged. Plants 1c and 2c were defined as outliers because their root Se-content was higher than their shoot Se-content. For all others, it was found that not only did pot-size play a role in the Se-uptake, but plant mass as well. Values for the two outliers 1c and 2c, as well as values for plant 3a and 5a representing 5 and 2.5 mg/L Se addition, respectively, are highlighted in each of the following graphs for better orientation.

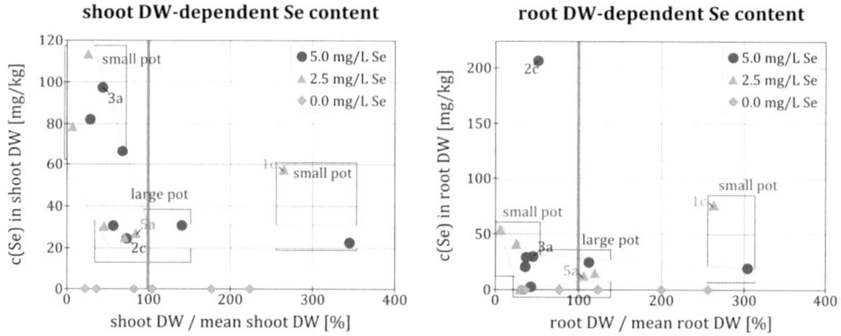

(a) (b)

Figure 3.10: Tissue Se content per percentage of average mass as an indicator of plant health
a: Shoot Se content per percentage of average shoot mass
b: Root Se content per percentage of average root mass

Figure 3.10a shows that plants with a lower-than-average shoot mass (7 - 44 %) – which were all found in small pots – showed the highest Se-shoot content (78 - 113 mg/kg DW). Selenium content decreased (67 - 24 mg/kg DW) as shoot mass approached average values (45 - 84 %), which was typical of plants from large pots. Selenium content remained low (31 - 23 mg/kg DW) in plants of higher-than-average mass grown in small pots, except for the outlier 1c (57 mg/kg DW for 265 % of average mass). For roots, disregarding the outliers 1c and 2c, this trend was similar, with plants of lower-than-average root mass (6 - 35 %)

reaching the highest Se content (54 - 21 %) and plants with root mass closer to average (29 - 45 %) reaching the lowest Se content (0 - 30 mg/kg DW). For plants with above-average root mass (106 - 120 %), Se content remained low (13 - 24 mg/kg DW). Pot-size classification of root-Se was similar to that of shoots.

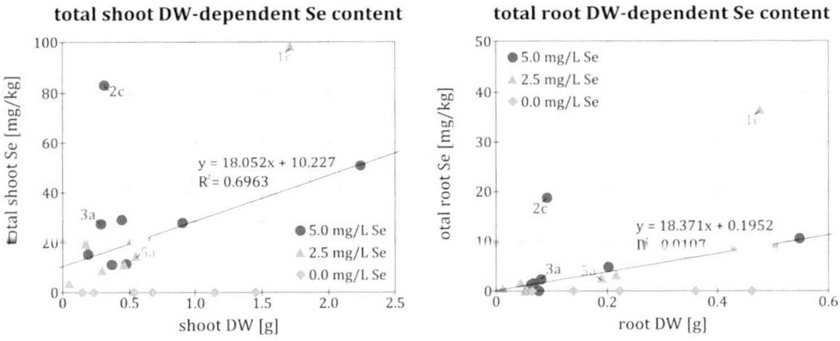

(a) (b)

Figure 3.11: Independence of Se-uptake of plant health
a: linear correlation between total shoot Se and shoot DW
b: linear correlation between total root Se and root DW

In cases in which Se-uptake is similar for two plants of different masses, pure plant mass could lead to dilution effects that have nothing to do with Se-uptake properties. Any non-linear correlation of Se content with plant mass is, therefore, a sign of either dilution effects or defective mechanisms in uptake regulation due to impaired plant health. Figure 3.11 shows there was no evidence of non-linear behavior. Instead, linear correlations between Se content and plant mass for both shoots and roots showed a good linear fitting ($R^2 = 0.6933$) for shoot-DW-dependent Se content and a very good fitting for root-DW-dependent Se content ($R^2 = 0.9197$) for both Se concentrations added.

3.5.2 Exp. A2 – plant-box rice cultures in Se-agar

Quality measures for experiment A2 (Chapter 3.3.2) showed that harvested plant yield was 84 % (±6 %) of planted caryopses for selenite and 88 % (±6 %) for selenate. Digestion standard retrieval was 84 % (±9 %) and 85 % (±19 %) for selenite and selenate, respectively, and drinking water standard retrieval during HG-FIAS measurement was 102 % (±7 %) for all samples of experiment A2.

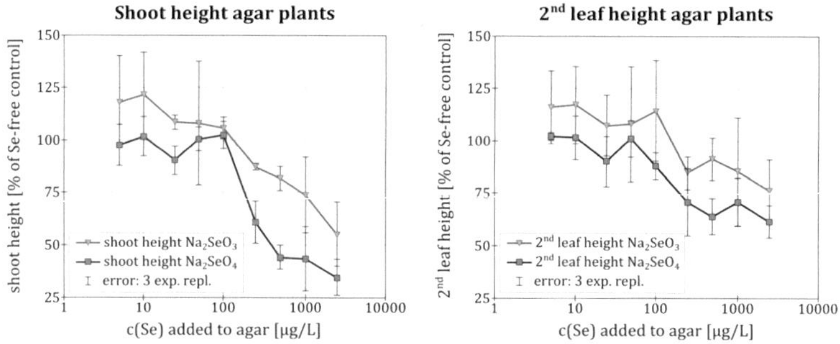

(a) (b)

Figure 3.12: Plant growth expressed as percentage of Se-free blank plants for experiment A2
a: total shoot height on harvest day (Day 16)
b: total length of 2nd leaf on harvest day (Day 16)

As shown in Figure 3.12a, total shoot height increased with the addition of Se as selenite between 0 and 10 µg/L, with 10.7 cm (122 %) being the maximum height reached, compared to 8.9 cm (100 %) height in the Se-free blank. If the concentration of selenite in the medium was 100 µg/L or higher, total shoot height decreased down to 5.0 cm when adding 2500 µg/L Se as selenite to the agar. Thus, selenite-treated plants reached only 55 % of the original plant height for 2500 µg/L Se. With the addition of Se as selenate, total shoot height remained similar for Se concentrations up to 100 µg/L, between 9.3 (90 %) and 10.5 cm (102 %), compared to the untreated plant height. Concentrations beyond that led to a sharp decrease in total shoot height, resulting in an overall decrease of 6.7 cm, from 10.2 cm (100 %) to 3.5 cm (35 %) when adding 2500 µg/L Se.

Similarly to total height, the length of the 2nd leaf (Figure 3.12b) was also affected by the same concentrations of selenite and selenate added to the agar. Low concentrations of selenite in the medium promoted growth (114 % for 100 µg/L Se), while concentrations above 100 µg/L were clearly inhibitory, with 76 % growth height for 2500 µg/L Se. Growth of the 2nd leaf was also negatively affected by selenate concentrations in the agar higher than 100 µg/L Se. At the highest selenate concentration, 2nd leaf height was 61 % of the untreated leaf height and the third leaf of the seedlings was almost entirely covered. Overall, selenite and selenate affected seedling growth similarly, with selenate being more effective.

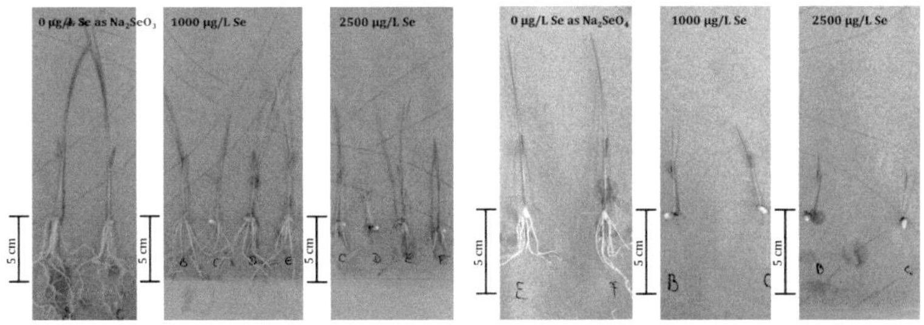

(a) (b)

Figure 3.13: Root growth and toxicity symptoms for the addition of 0, 500, 1000 and 2500 µg/L Se to agar as
a: selenite
b: selenate

Table 3.4: Calculated absolute Se uptake (Se as Na_2SeO_4) by plants from agar Se-content of Exp. A2 using HG-FIAS measurement data

Se added [µg/L]	initial agar c(Se)FIAS [µg/L]	initial agar inweight [g]	initial agar c(Se)agar [µg/L]	final agar c(Se)FIAS [µg/L]	final agar inweight [g]	final agar c(Se)agar [µg/L]	uptake/plant box Δc(Se) [µg/L]	uptake/plant box absol. Se [µg]
0	< PQL	1.5558	< PQL	< PQL	4.3209	< PQL	< PQL	< PQL
5	< PQL	1.6793	< PQL	< PQL	4.3049	< PQL	< PQL	< PQL
10	< PQL	1.2469	< PQL	< PQL	4.3184	< PQL	< PQL	< PQL
25	< PQL	1.9685	< PQL	< PQL	4.2925	< PQL	< PQL	< PQL
50	7.29	1.6599	43.90	4.06	4.3150	9.40	34.50	0.15
100	15.95	1.7854	89.33	7.78	4.3120	18.04	71.29	0.31
250	43.01	2.1887	196.51	19.10	4.2611	44.83	151.68	0.65
500	31.82	1.1443	278.08	64.75	4.3154	150.04	128.04	0.55
1000	134.39	1.4243	943.55	132.78	4.2699	310.96	632.59	2.70
2500	344.93	1.5222	2265.97	396.48	4.3136	919.13	1346.84	5.81

Root lengths remained similar at 5 - 7 cm for the addition of 0 - 100 µg/L Se for both Se species (Figure 3.13). After that, root length decreased steadily to below 1 cm when Se was added as selenate. Additional selenite, however, resulted in decreased root length (1 - 2 cm) only for 2500 mg/L Se. Toxicity symptoms of stunted root growth, lack of secondary roots and brown-spotted discoloration were found for the addition of 250 to 2500 µg/L Se as selenate (Figure 3.13b), but only for 2500 µg/L Se as selenite (Figure 3.13a).

Unfortunately, measurements of Se content in agar by acid digestion were not precise enough to be significant. No Se could be detected for additions of 25 µg/L Se or lower, which indicates a method error of at least 25 µg/L, which is insufficient for the determination of plant uptake from the agar via Se loss from 100 mL of agar. Values calculated for absolute Se uptake (Table 3.4) are, therefore, not considered trustworthy and absolute Se uptake could only be determined for plants of Exp. A3 (Chapter 3.3.3), because the nutrient solution could be directly measured by ICP-MS.

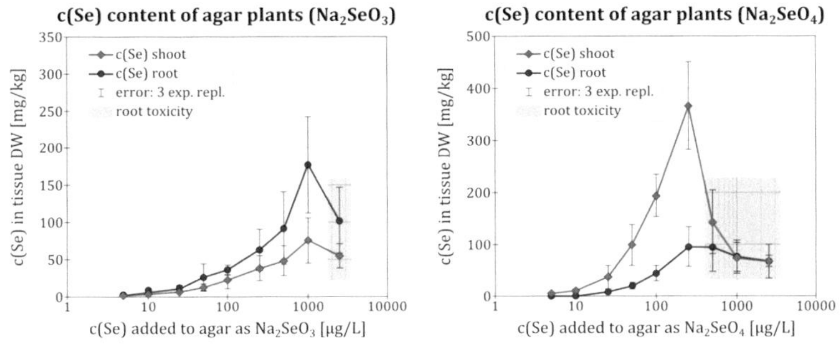

(a) (b)

Figure 3.14: Se content in roots and shoots of experiment A2
a: addition of Se as Na_2SeO_3
b: addition of Se as Na_2SeO_4

As shown in Figure 3.14a, when adding selenite to agar at a concentration of 1000 µg/L, Se concentrations in both roots and shoots steadily increased to 177 mg/kg Se plant dry weight in the roots and 75 mg/kg Se DW in the shoots. If selenite concentration was increased further (i.e. 2500 µg/L), Se content in roots (102 mg/kg Se DW) and shoots (55 mg/kg Se DW) decreased, which is consistent with toxicity symptoms found only for 2500 µg/L Se. Of all Se found within the plant when adding it as selenite, 30 - 38 % of the Se was found in the shoots, while 62 - 80 % was found in the roots (Table 3.5). With the addition of Se as selenate, Se concentrations in shoots were generally greater than in roots (Figure 3.14b). Selenium concentrations for both roots and shoots increased with additional Se to the agar until a maximum of plant-Se was reached at the addition of 250 µg/L Se for shoots (367 mg/kg Se DW) and roots (96 mg/kg Se DW). In the shoots, additional Se led to a decrease in plant-Se to 143 mg/kg Se DW for 500 µg/L Se until reaching a plant-Se value of 67 mg/kg Se DW when plants were raised in 2500 µg/L Se. This was below the Se concentration of 98 mg/kg Se

DW found in the plant when adding 50 µg/L Se. In the roots, a decrease of plant Se concentration was found if the added concentration of Se exceeded 1000 µg/L. This is also consistent with root toxicity symptoms found at added Se concentrations between 500 and 2500 µg/L Se. Of all Se found within the plant when adding it as selenate, 50 - 82 % of the Se was found in the shoots, while 18 - 50 % was found in the roots (Table 3.5).

Table 3.5: Se distribution for uptake of selenate and selenite into root and shoot in both experimental set-ups A2 and A3 (d_{shoot} [%] = $c(Se)_{shoot}$ [mg/kg]/($c(Se)_{shoot}$ [mg/kg] + $c(Se)_{root}$ [mg/kg]) · 100), standard deviation (SD) given as 1s

	agar experiments						**nutrient solution experiments**					
	Na_2SeO_3			Na_2SeO_4			Na_2SeO_3			Na_2SeO_4		
added c(Se) [µg/L]	shoot [%]	root [%]	SD [1s]	shoot [%]	root [%]	SD [1s]	shoot [%]	root [%]	SD [1s]	shoot [%]	root [%]	SD [1s]
5	38	62	7	77	23	6	28	72	8	62	38	28
10	32	68	7	76	24	8	33	67	1	78	22	19
25	37	63	7	81	19	11	30	70	8	74	26	21
50	32	68	10	82	18	7	20	80	11	70	30	1
100	38	62	16	81	19	9	26	74	12	73	27	7
250	38	62	8	79	21	11	34	66	3	70	30	10
500	34	66	3	60	40	5	30	70	3	73	27	9
1000	30	70	8	49	51	16	36	64	3	75	25	6
2500	35	65	6	50	50	16	34	66	3	76	24	8

3.5.3 Exp. A3 – plant-box rice cultures in Se-nutrient solution

Quality measures for experiment A3 (Chapter 3.3.3) showed that harvested plant yield was 71 % (±9 %) of planted caryopses for selenite and 63 % (±6 %) for selenate. Digestion standard retrieval was 89 % (±13 %) and 84 % (±7 %) for selenite and selenate, respectively, and drinking water standard retrieval during HG-FIAS measurement was 104 % (±5 %) for all samples of experiment A3.

Figure 3.15a shows that for the nutrient solution experiments, the addition of 2500 µg/L Se as selenite resulted in an apparent decrease of 5 cm from 24 cm to 19 cm of the total shoot height, which equalled 79 % of the height of plants grown in the absence of additional Se. This overall decrease, however, was still within 1s-error margin, even when normalized to untreated plant height. This lack of trend was also reflected in the 2nd leaf heights (Figure 3.15b), which remained constant between 93 % and 104 % of untreated leaf height.

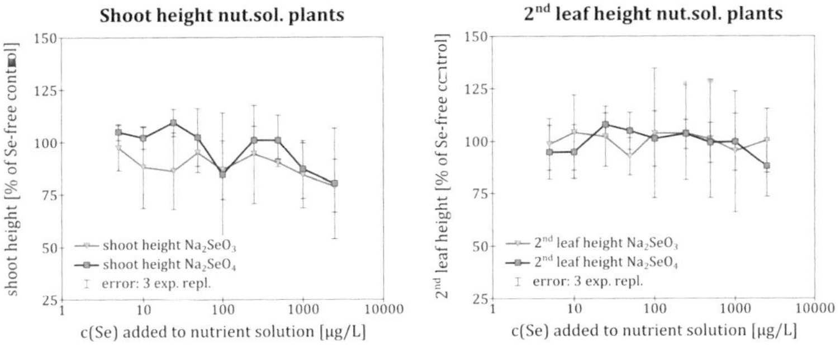

(a) (b)

Figure 3.15: Plant growth expressed as a percentage of Se-free blank plants for experiment A3
a: total shoot height on harvest day (Day 19)
b: total length of 2nd leaf on harvest day (Day 19)

Similarly, total shoot heights showed an overall decrease of 5 cm (from 23 cm to 18 cm) with the addition of selenate into the nutrient solution (Figure 3.15 a). Although this equalled 80 % of the untreated plant height, this decrease was also still within the 1s-error margin when normalized to untreated plant height and 2nd leaf heights (Figure 3.15 b) remained between 108 % and 88 % of untreated leaf height. This showed that – unlike the results in the agar experiments – neither selenite nor selenate addition had any significant effect on plant growth in nutrient solution experiments. The lack of root toxicity symptoms in nutrient solution experiments, which, in contrast, were pronounced in the agar experiments, confirmed this. A comparison of shoot height of the untreated plants on Day 9 between both experiments showed that agar plant shoots grew to a height of 3.4 cm (±0.8), while nutrient solution plants grew to a height of 6.6 cm (±2.5) when no Se was present.

As shown in Figure 3.16a, increasing the addition of Se as selenite to the nutrient solution resulted in an almost linear increase in Se concentration in both shoots and roots. Selenium concentrations reached 312 mg/kg Se DW in roots and 159 mg/kg Se DW in shoots for the Se-addition of 2500 µg/L Se. This pattern of uptake differed from the agar experiment uptake, where an uptake peak was observed at a concentration of 100 µg/L of added Se, but additional Se led to lower plant-Se. Maximum uptake into the rice plant in the agar experiments was, therefore, lower than in the nutrient experiments. Figure 3.16b shows a similar linear increase for plant-Se in plants grown in nutrient solution with added selenate. Here, Se concentrations reached 405 mg/kg Se DW in the shoots and 128 mg/kg Se DW in the roots.

Again, an uptake peak and subsequent decrease, similar to the agar experiments, was not observed and maximum uptake was higher than in agar experiments. Furthermore, significant root length differences between added Se concentrations were not found, irrespective of the Se species added to the medium. Error (error bars over all three experimental runs, Figure 3.16) decreased with the addition of Se in Exp. A3, with an error over all plant-Se of 48 % (as Na_2SeO_3 and Na_2SeO_4) with the addition of 10 - 50 µg/L Se, 30 % with the addition of 100 - 500 µg/L Se and 20 % with the addition of 1000 - 2500 µg/L Se.

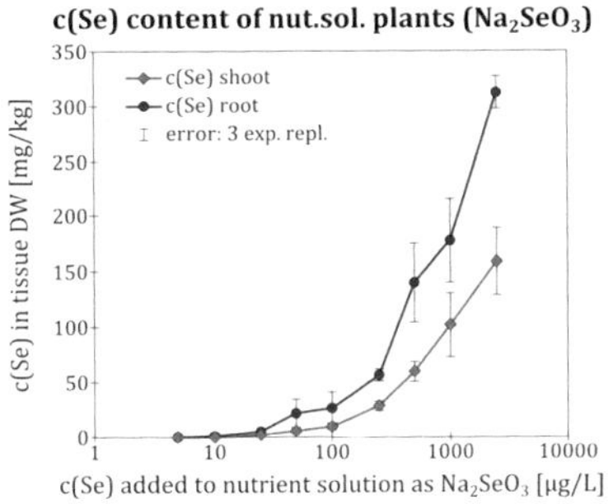

(a)

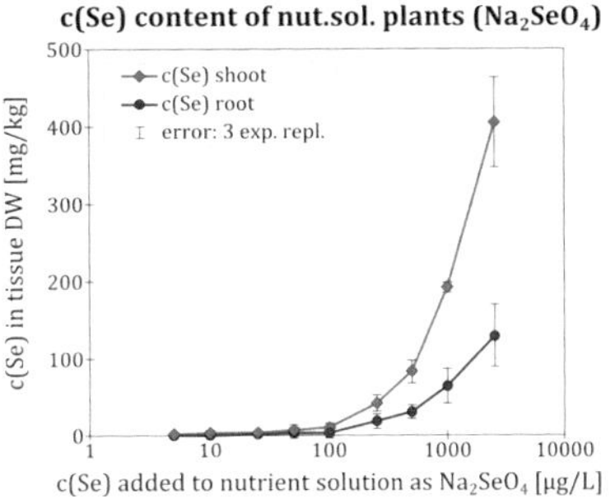

(b)

Figure 3.16: Se content in roots and shoots of experiment A3
a: addition of Se as Na_2SeO_3
b: addition of Se as Na_2SeO_4

Selenium accumulation factors (Table 3.6) of fresh plant weight showed that on average, accumulation of Se as selenite in the roots was 2.7 (±0.2) times higher than in the shoots for Exp. A2 and 3.3 times higher (±0.4 when excluding the outlier at 50 µg/L Se) than in the shoots in Exp. A3. In contrast, Se added as selenate led to an accumulation factor of 2.8 (±0.2 when excluding the toxicity outliers) and 1.9 (±0.3) times higher in the shoots than in roots in Exp. A2 and A3, respectively. For the addition of 1000 and 2500 µg/L Se as selenate to agar, shoot accumulation was lower than in roots (factor 1.5). Accumulation of Se in plant tissue differed between Exp. A2 and Exp. A3. Selenium present as selenite in Exp. A2 showed an overall decrease in accumulation factors with higher concentrations of initial Se from 75 - 7.2 and 198 - 18 for the addition of 5 - 2500 µg/L Se for shoots and roots, respectively (Table 3.6). When present as selenate, Se accumulation in the plant tissue in Exp. A2 showed an increase from 274 - 514 and 108 - 175 for the addition of 5 - 100 µg/L Se for shoots and roots, respectively. Addition of Se in concentrations between 100 and 2500 µg/L Se, however, led to a decrease in Se accumulation in the plant tissue from 514 to 8.4 and 175 to 13 for shoots and roots, respectively (Table 3.6). Selenium accumulation in Exp. A3 was similar across a large range of initial Se concentrations with 22 - 34 and 80 - 117 for shoots and roots, respectively, when Se was added as selenite and 19 - 41 and 9.7 - 23 for shoots and roots, respectively, when Se was added as selenate.

Data on the uptake of nutrients by the plant showed no significant trends when plotted as a function of added Se. Neither selenite nor selenate had a measurable effect on nutrient uptake of N, P, K, S, Ca, Mg, Mn, Zn, Cu or Fe, as all fluctuations were well within error range of the 3 runs of the experiment. Two elements, Na and Cl, were not included because, due to climate chamber humidity operation, Na and Cl data could not be interpreted. This is because their presence in the water, which was used to maintain humidity in the climate chamber, led to an increase in Na and Cl content in the nutrient solution over the time period of the experiment, rather than a decrease due to plant uptake. Furthermore, B was found to be leached from lab equipment and the plant-boxes, leading to greater nutrient solution content of B after the

Table 3.6: Fresh weight accumulation factors (AF) of selenate and selenite into shoots in both experimental set-ups (AF [-] = $c(Se)_{medium}$ [mg/L] / $c(Se)_{plant}$ [mg/kg])

	agar experiments				**nutrient solution experiments**			
	AF-Na_2SeO_3		AF-Na_2SeO_4		AF-Na_2SeO_3		AF-Na_2SeO_4	
added c(Se)	shoot	root	shoot	root	shoot	root	shoot	root
[µg/L]	[%]	[%]	[%]	[%]	[%]	[%]	[%]	[%]
5	75.35	198.12	273.80	108.36	21.54	80.31	27.22	21.71
10	53.15	134.48	263.10	99.13	26.67	81.57	16.20	8.03
25	59.27	129.49	390.10	130.07	26.47	99.16	20.88	9.68
50	50.37	145.38	465.25	154.76	31.59	211.88	26.74	12.68
100	52.89	146.66	513.57	175.22	26.97	104.61	19.00	10.20
250	33.14	92.40	366.44	132.03	28.98	94.10	41.13	19.86
500	28.14	76.26	93.92	79.82	33.61	116.47	39.99	21.66
1000	22.98	68.02	25.19	37.00	30.68	84.81	53.35	25.53
2500	7.22	17.68	8.36	13.09	22.15	57.82	40.87	22.51

experiment than initially supplied. It appears that with the addition of selenate, plant uptake of Mn and Zn showed a slightly decreasing trend at concentrations of 1000 and 2500 µg/L. However, all nutrient solution phases including Fe are meta-stable from a thermodynamic point of view (as calculated with Phreeq-C), which makes co-precipitation of such small amounts of Mn and Zn that were added to the solution highly probable. Therefore, data on nutrient uptake is considered uncorrelated with addition of Se in all measured cases.

3.5.4 Exp. A4 – plant-box rice cultures in Se-agar with delayed Se-uptake

Quality measures for experiment A4 (Chapter 3.3.4) showed that harvested plant yield was 48 % (±13 %) of planted caryopses for selenite and 72 % (±21 %) for selenate. Digestion standard retrieval was 101 % (±4 %) and 95 % (±7 %) for selenite and selenate, respectively, and drinking water standard retrieval during HG-FIAS measurement was 118 % (±7 %) for all samples of experiment A4.

Although there was no standard deviation for experimental data of experiment A4 because there were no repeats, conditions were similar to experiments A2 and A3, which is why results are also given as percentages of untreated plant growth for comparison.

With this in mind, Figure 3.17a still exhibited a trend in total shoot heights with the addition of 500 µg/L Se or more to the agar from 17 cm (100 %) to 10 cm (59 %) for selenite and 19 cm (100 %) to 8 cm (41 %) for selenate. Additions of 250 µg/L Se or less showed no change in shoot height for either speciation. Second leaf heights, as shown in Figure 3.17b, however, showed no significant decrease compared to untreated or low-Se leaf heights.

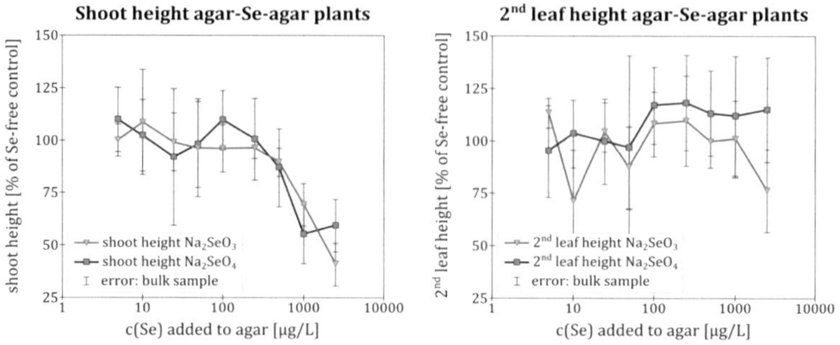

(a) (b)

Figure 3.17: Plant growth expressed as percentage of Se-free blank plants for experiment A4 (note: error bars refer to single plant deviations from bulk mean)
a: total shoot height on harvest day (Day 19)
b: total length of 2nd leaf on harvest day (Day 19)

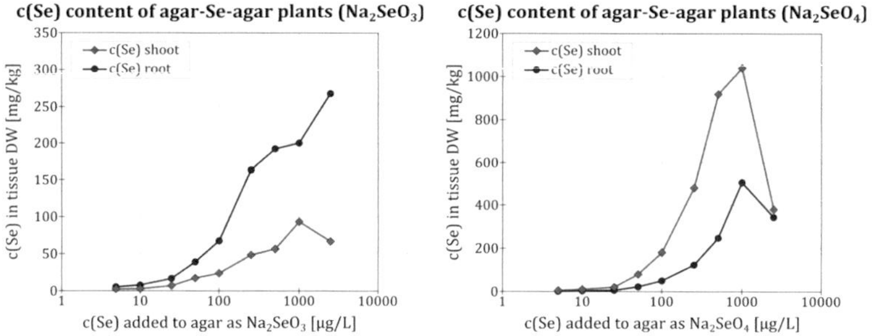

(a) (b)

Figure 3.18: Se content in roots and shoots of experiment A4
a: addition of Se as Na_2SeO_3
b: addition of Se as Na_2SeO_4

When Se was applied as selenite (Figure 3.18a), Se content in roots increased sharply between 0 and 250 µg/L of added Se from 0 to 164 mg/kg DW. Selenium content increased less strongly with the addition of more than 250 µg/L Se, resulting in 269 mg/kg DW Se with the addition of 2500 µg/L Se. In shoots, Se content peaked with the addition of

1000 µg/L Se as selenite at 93 mg/kg DW and decreased to 67 mg/kg with the addition of 2500 µg/L. As shown in Figure 3.18b, Se applied as selenate resulted in increased Se content with increasing Se addition until peaking in both shoots (1040 mg/kg DW) and roots (505 mg/kg DW) with the addition of 1000 µg/L Se. A higher Se concentration of 2500 µg/L in the agar led to a decrease in Se content in shoots (382 mg/kg DW) and roots (345 mg/kg DW).

Because two plants in the 2500 µg/L plant-box of the selenite experiment were unusually small and exhibited discoloring around the caryopse even though their roots had not reached the Se-agar, these were sampled separately. Together with agar samples from the top 2 cm of the non-Se-bearing Eppendorf tube, these were separately tested for stray traces of Se. Agar digestion as well as plant digestion of these samples revealed Se-content to be below detection limit. Therefore, the 0.7 % agar in the Eppendorf tubes can definitely be considered Se-free at plant germination.

When directly compared to results from experiments A2 and A3, experimental data of A4 for selenite (Figure 3.19) exhibited three characteristics. The first was that for uptake of selenite, data on shoot-Se followed the trend observed in results of experiment A2, with a decrease in shoot Se-content at concentrations greater than 1000 µg/L (Figure 3.19a). While the trend was similar, uptake concentrations for experiment A4 were between 17 % and 26 % higher than in the A2 experiment. However, as a second characteristic, for uptake of selenite (Figure 3.19b), data on root-Se followed the trend of experiment A3, in which there was no collapse in Se-uptake at concentrations greater than 1000 µg/L. The increase in uptake between the addition of 1000 and 2500 µg/L Se, however, was only 25 %, compared to 45 % in experiment A3.

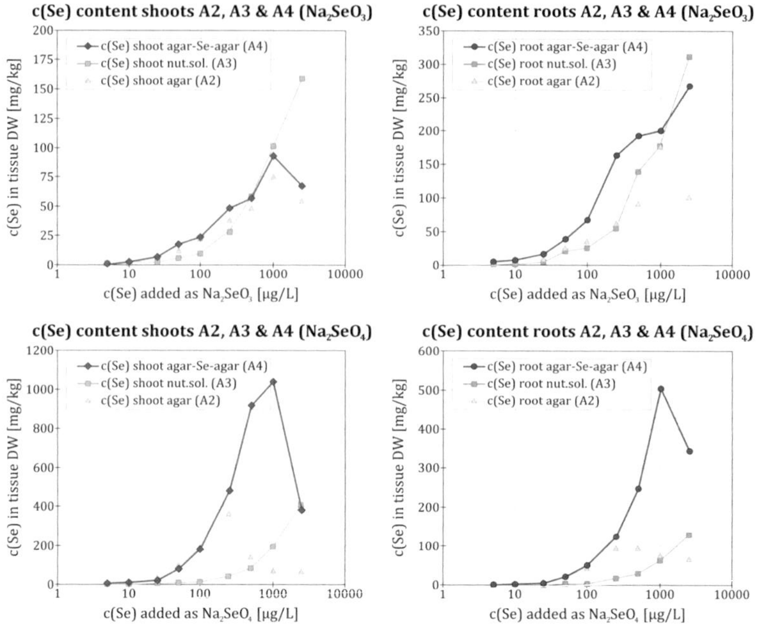

(a) (b)

Figure 3.19: Comparison of Se content of experiment A2, A3 and A4 after uptake as Na_2SeO_3 (top) and Na_2SeO_4 (bottom) (error bars for A2 and A3 not included for better visibility)
a: shoot-Se for A2, A3 and A4
b: root-Se for A2, A3 and A4

As a third characteristic, root-Se after uptake of selenite also showed a similarly strong increase with additions of Se between 100 and 1000 µg/L Se, as is exhibited in experiment A2 (Figure 3.19b). This allowed the conclusion that for selenite, the set-up of experiment A4, which was designed to explore the parameter of nutrient deficiency of experiment A2 combined with the delayed uptake of Se of experiment A3, reflected this intermediate set-up in the results.

For the addition of selenate, comparison of results of experiments A2, A3 and A4 reflected the intermediate set-up of experiment A4 as well, but in a different way than was the case for selenite Figure (3.19). For both shoots and roots, plant-Se was nearly identical as for experiments A2 for added concentrations of Se between 0 and 100 µg/L. After that, however, the trend of increasing uptake was continued similarly to the trend observed in results of experiment A3 - albeit at much higher concentrations. For the addition of 1000 µg/L, Se content was about 5 and 8 times higher in shoots (Figure 3.19a) and roots (Figure 3.19b) in experiment A4 than in experiment A3, respectively. At concentrations exceeding 1000 µg/L Se, Se content decreased sharply, following the trend for Se-uptake in experiment A2, albeit at higher additions of Se and higher plant-Se content.

3.6 Discussion on the uptake of Se into plants

The following subchapter will discuss how plant status, nutrient supply, Se-exposure at germination and Se speciation result in different plant Se-uptake and distribution in experiments A1, A2, A3 and A4. The main conclusions are derived from experiments A2 and A3, with A1 and A4 providing supporting results.

3.6.1 Influences of plant status and substrate amount on Se-uptake in rice

The greatest differences in the Se-uptake of the open rice cultures were not – as anticipated – between the two different concentrations of 2.5 or 5 mg/L, or even between the amounts of Se-nutrient solution supply, but between pot size. This is interesting, as pot size not only changed the ratio of substrate to solution and therefore diluted the applied amount of Se (which simultaneously also applies to the nutrients of the solution). Pot size also changed the ratio of root surface area to substrate. Therefore, plants in the small pots and, therefore, less substrate were more likely to come into contact with more Se, which was verified by plant-Se results.

Pot size-related substrate amount apparently played a role in plant development as well (Figure 3.9). Especially when comparing shoot mass, the larger substrate amount produced plants that were more homogenous and average in development compared to the plant with smaller substrate amounts, which, in turn, reflected Se-uptake. The fact that plants with the lowest-to-average plant mass took up the highest Se concentrations, while plants with average and above-average plant mass plateaued in their Se content, indicates an interesting correlation between uptake and low plant health. Therefore, the hypothesis that plants with less vitality due to less watering and nutrient sustenance would be affected more by the addition of Se than plants with greater vitality was supported. Plotting the absolute Se in the plant against shoot mass as an indicator of plant health was an attempt at clarification of the cause-and-effect relationship. However, a clear separation of cause and effect between plant health and Se toxicity for the open cultures was not possible due to low sample numbers and further discussion will focus on the following experiments, which explore nutrient supply and toxicity more closely.

3.6.2 The influence of nutrient supply on Se-uptake

For application of lower Se concentrations (< 250 µg/L as selenate), Se-uptake is much higher in the nutrient-free environment (ca. factor 12) compared to nutrient-complemented plants, a conclusion that is also supported by substantially higher (5 - 27 times) accumulation factors (Table 3.6). This conclusion is consistent with published findings [Buchner, Takahashi & Hawkesford, 2004, Li et al., 2007] that depletion of sulfate and phosphate increases the activity of the respective transporters, so that more Se is taken up into the plant. Furthermore, lack of competing ions, i.e. sulfate, is expected to further promote the efficiency of selenate uptake. This is most prominently demonstrated in the nutrient-free experiment with a delay before Se-uptake, where Se uptake for selenate by far exceeds the highest uptakes of the other experimental set-ups due to lack of nutrients (factor 2.5 and 4 for shoots and roots, respectively). However, when a threshold of Se toxicity is exceeded (250 µg/L Se as selenate and 1000 µg/L as selenite in experiment A2 and 1000 µg/L as selenate in experiment A4), Se-content in the plant collapses. For plants of experiment A2 this also coincides with root toxicity symptoms, which indicates that root functionality is impaired and therefore transporter protein activity is shut down, significantly lowering Se-uptake.

However, addition of nutrients reduced Se-uptake when Se concentrations were < 500 µg/L, which was especially evident for selenate (Figure 3.16). This is also reflected by the accumulation factors – all of which are lower than in their nutrient-free counterparts (Table 3.6). This might be caused by down-regulation of transporter activity in response to sufficient content of nutrients (i.e. S), or by competition of sulfate for binding sites at the transporters [Li et al., 2007, Khan & Hell, 2014]. Even for the highest tested concentration of Se (2500 µg/L Se, corresponding to approximately 30 µmol/L Se as SeO_4^{2-} or SeO_3^{2-}), the 700 µmol/L SO_4^{2-} and 400 µmol/L PO_4^{3-} of the nutrient solution are most likely to compete severely with Se for uptake into the plant. It is known that additional nutrient oxy-anions, especially phosphate and sulfate, influence Se-uptake by plant roots [Li et al., 2007] [Khan & Hell, 2014]. Effects of S are considered to be more pronounced than those of P, because the affinity of selenate for the sulphur transporter is higher than that of selenite for the phosphate transporter [Hopper & Parker, 1999, Li et al., 2007].

For Se concentrations above the nutrient-free toxicity-threshold (1000 µg/L Se as selenite and 250 µg/L Se as selenate), nutrients clearly promoted the uptake of Se. This correlates with the fact that the nutrient-complemented plants remained healthy even when exposed to high Se concentrations and thus, presumably, were capable of sustaining uptake, leading to a higher Se-uptake compared to the nutrient-free condition. This is consistent with a previous study showing that higher S content has a protective physiological effect on Se toxicity [Kikkert et al., 2013]. This protective effect of nutrition is agronomically relevant, because at 2 mg/kg DW Se, rice yield is already reduced by 10 % and this yield loss is even more pronounced if the Se is administered as selenite [Läuchli, 1993]. Therefore, application of nutrients could help lower Se toxicity and improve crop yield in general. Although Se-uptake in our experiments was facilitated because the Se was freely available in a solution rather than soil-bound, we did not observe any drop in uptake nor any significant inhibition of plant growth, even for a condition yielding 405 mg/kg Se DW in the shoots. Thus, our study suggests that plant resilience increased with nutrient availability. Moreover, competition between anions such as sulfate and selenate as well as selenite and phosphate not only affected the uptake, but also the partitioning between shoots and roots (Chapter 3.6.4). This indicates that sulfate competition for selenate uptake might also influence the transport from roots to shoots and/or the incorporation of Se into proteins, in accordance with previous findings [Li et al., 2007].

3.6.3 Phytotoxicity triggered by direct Se-exposure and lack of nutrients

Of all three closed-box experiments, toxicity symptoms on plants were only observed for plants grown in the nutrient-free environment, in which seedlings were in direct Se contact upon germination. As plants that showed no toxicity symptoms all came into contact with Se after a delay of about 5 days, this strongly supports the hypothesis of toxicity symptoms being triggered, though not exclusively caused, by direct Se exposure.

So far, effects of Se toxicity or biofortification have only been tested on healthy, pre-germinated seedlings [Asher et al., 1977, Bellet al., 1992, Zhang et al., 2006, Li et al., 2007]. Only one study directly addressed germination of rice in the presence of Se [Liu & Gu, 2009]. However, seedling development beyond the radicle piercing the seed coat was not investigated. In that study, germination itself was not influenced significantly below 60 µmol/L (ca. 4.74 mg/L) Se in solution, in contrast to the much lower concentration of 1 - 2.5 mg/L Se observed to impair growth in our study. This suggests that germination per se is not the most sensitive step but rather, the subsequent development of the germinated seedling, especially under conditions of nutrient deficiency. Germination of rice in the presence of Se has been proposed as a strategy of effective Se biofortification [Liu & Gu, 2009].

Experiments of this study provide strong evidence that it is the combination of a lack of nutrients, coupled with direct Se exposure above the threshold concentrations (250 µg/L Se as Na_2SeO_4 and 1000 µg/L Se as as Na_2SeO_3), that leads to toxicity symptoms. Nutrient deficiency alone, although leading to smaller plants, did not lead to toxicity – neither in the agar blanks of the nutrient-free experiments with germination directly in Se, nor in the agar blanks of the nutrient-free experiments with Se. Even the combination of nutrient deficiency and a delay between Se uptake and germination (as in experiment A4) did not lead to the toxicity symptoms observed in the experiments, in which seedlings germinated in Se exposure. Therefore, this study suggests that it is essential for the strategy of Se biofortification at the pre-germination stage to provide sufficient nutrients to young seedling to avoid poisoning seedlings and lowering crop yield even at low Se concentrations. The alternative strategy might be to add Se at a later growth stage, or to pre-cultivate seedlings in a Se-free environment, before subjecting them to high Se concentrations after 5 days.

3.6.4 Selenium-uptake partitioning – distribution of selenate and selenite uptake into shoots and roots

The second main observation was that the addition of Se as selenate resulted in Se accumulating predominantly in the shoots rather than the roots, whereas addition of selenite resulted in Se accumulating mainly in the roots rather than in the shoots. This was the case for all plant-box experiments (Figures 3.14, 3.16, 3.18, Table 3.5). Mean Se concentrations in shoots were up to 3.5 - 5 times higher than in roots when Se was applied as selenate. In contrast, when applied as selenite, Se concentrations were up to 2 - 4 times higher in roots compared to shoots. This inverse partitioning was independent of the mode of cultivation.

The preferential partitioning of selenate into shoots has also been reported for other plants, such as strawberry, clover and perennial ryegrass [Hopper & Parker, 1999], barley and red clover [Gissel-Nielsen, 1973], soybeans [Zhang et al., 2003] and wheat [Li et al., 2007]. This phenomenon is generally explained by selenate being taken up actively via the sulfate transporter and then readily transported via xylem into plant shoots and leaves [Carey et al., 2012]. However, the behaviour of selenite uptake is not fully understood [Khan & Hell, 2014]; selenite uptake mechanisms, for instance, are still a matter of debate.

These experiments showed that plant Se-content is nearly as high for the uptake of selenite compared to selenate, which is known to be mediated by high-affinity transporters. Furthermore, accumulation factors (Table 3.6) of selenite in roots are as high or higher than accumulation factors of selenate in shoots.

These findings support a model in which selenite is taken up actively via phosphate transporters against the electrochemical gradient as previously suggested [Hopper & Parker, 1999, Li et al., 2007]. The selenite taken up in both of our experimental set-ups – presumably by phosphate [Li et al., 2007] or silicon transporters [Zhao et al., 2010] – is likely to be converted into organic seleno-compounds. This was also shown by speciation analysis of plant material in this study (Chapter 5.4.3).

On the one hand, organic Se can be methylated to produce the non-proteinogenic amino acid selenomethylcysteine [Khan & Hell, 2014]. This amino acid has been shown to accumulate in chloroplasts of *Arabidopsis*, which appear to use specific methylation as a mechanism to prevent mis-incorporation of the Se-amino acid into proteins instead of S-amino acids [LeDuc et al., 2004].

On the other hand, it was found that selenomethionine (SeMet) and selenomethylselenocysteine (SeMeSeCys) were rapidly transported exclusively via the phloem to the grain, whereas selenate was transported slower via the xylem [Carey et al., 2012]. This poses the question whether that one third of selenite, which was transported to the shoots is incorporated into proteins or retained in shoot chloroplasts. It appears that rice plants harbour protective mechanisms similar to those of Se-accumulator plants to retain two-thirds of selenite as methylated compounds in the root that are not integrated in proteins. It remains to be investigated whether this apparently precise partitioning of selenite is caused passively by saturation of binding sites, or is actively regulated.

The fact that most of the selenate is partitioned to the shoots as a consequence of its unspecific transport via sulfate transporters explains the large share of selenate in the shoot. However, it does appear that a third of the selenate taken up by plant is retained in the root – most likely as organic Se as it can be reduced to selenide and then transformed into SeCys [LeDuc et al., 2004, Khan & Hell, 2014].

3.6.5 Conclusions for biogeochemical mass balance modelling

Results of this study can be implemented into the biogeochemical model (Chapter 6) for the characterization and mass balance of the Se compartments and transfer processes within the Critical Zone. Mathematical description of uptake as functions of Se concentration present in a certain species is, of course, merely a quantification of transport, and is not descriptive of any single chemical or physical property or process. Therefore, the biogeochemical model can only provide adequate description of correlations and quantification of processes.

Plant uptake processes describe one half of the predefined model transport pathways for Se and applies to the compartments of 'soil solution' and 'plant'. Conclusions from this study for this aspect of the Critical Zone are:

1. selenite and selenate are taken up readily by the rice plant, but resulting Se-accumulation within the plant tissue is partitioned differently to root and shoot;

2. nutrient supply provides competition for Se-uptake, but the thereby optimally secured plant health leads to greater tolerance and ultimately higher Se-uptake for high Se concentrations;

3. direct exposure to Se during plant germination leads to phytotoxicity at moderate Se concentrations, but effective Se-uptake by the plant is increased at low Se concentrations.

4 Geochemistry – Se sorption processes onto kaolinite & goethite

The second part of this study is concerned with the interactions between soluble Se and soil-forming minerals, particularly adsorption and desorption. Two of the most frequent soil-forming mineral groups which play an important role in controlling Se mobility in soils are clay minerals and iron oxides or -hydroxides [Bar-Yosef & Meek, 1987, Ghose et al., 2010]. This study focuses on one representative mineral for each of these groups:

(1) as pure a kaolinite as could be obtained to represent a highly crystalline 2:1 layer clay mineral and

(2) an α-FeOOH (goethite), which in this case has a large share of amorphous properties.

4.1 Selenium retention in soils

Although underlying geology is one of the primary controls of soil-Se concentration due to weathering, mechanisms of soil retention determine bioavailability of the Se present in the soil [Fordyce, 2013]. During weathering, Se is readily oxidized and becomes more mobile with increased oxidation state [Neal, 1995]. Generally, in arid alkaline environments, Se is, therefore, most likely to be present as selenate, which is highly mobile, while humid regions promote the formation of selenite, which is less mobile [Neal, 1995]. At the interface between soil solution and solid phase, defining biophysical parameters such as soil mineralogy, texture and organic matter, influence and are influenced by chemical factors such as pH, redox conditions and competing ions [Fordyce, 2013].

4.1.1 Selenium associations in soil

Selenium is known to be associated with iron, aluminum and manganese oxides and hydroxides [Balistrieri & Chao, 1987, Hayes et al., 1987, Balistrieri & Chao, 1990, Chan et al., 2009, Das et al., 2013], carbonates [Wijnja & Schulthess, 2002], clay minerals [Bar-Yosef & Meek, 1987, Goldberg et al., 2013] and organic matter [Kang et al., 1991, Gustafsson & Johnsson, 1992]. Adsorption of Se by clay minerals can lower Se bioavailability to as little as 50 % and iron oxides can completely adsorb Se [Fordyce, 2013]. Maximum adsorption is influenced strongly by pH and reaches its maximum at pH 3 - 5, decreasing with increasing pH [Neal, 1995, Fordyce, 2013]. Chemical extraction procedures have shown that, depending on soil composition, Se adsorbed onto mineral phases accounts for about 20 % of the total soil Se in soils and sediments [Fernández-Martínez & Charlet, 2009], while an estimated 50 % of Se in soils and sediments are retained by organic matter [Fernández-Martínez & Charlet, 2009]. Fixation by organometallic complexes is the reason why organic matter can also significantly lower Se bioavailability to as little as 10 % [Neal, 1995, Fordyce, 2013]. In anoxic conditions, however, it was found that elemental Se can account for 20 - 60 % of the total Se in soils and sediments [Zhang & Moore, 1996, Fernández-Martínez & Charlet, 2009].

All these general influences are helpful to asess Se bioavailability in soils. However, when studying depth profiles of Se concentration in seleniferous soils of the Kesterson reservoir [Tokunaga et al., 1994] or in Punjab, India [Dhillon & Dhillon, 2003], where Se is enriched 5 - 10-fold in the top 15 - 40 cm compared to Se content near bedrock, it becomes apparent that mechanisms of Se soil retention and Se transport affect Se soil concentrations across greater areas. Furthermore, in natural soils, generalizations of environmental redox conditions do not account for the small-scale heterogenous redox conditions within soil aggregates. These microsites of anaerobic conditions within an environment of overall aerobic conditions are a product of slow oxygen diffusion rates and microbiological redox processes and can lead to a great variety of Se oxidation states across the scale of around 1 mm [Tokunaga et al., 1994]. For these reasons, soils are very difficult to model in their entire complexity.

Therefore, certain aspects of natural soils were not included in this study to simplify the set-up. Since this study focuses on inorganic soil properties, the significant impact of microbiology [Neal, 1995] is not detailed here and was not included. Furthermore, the aim of this study was to provide quantification of Se adsorption and desorption processes of soil-relevant minerals specifically tailored to complement the Se-plant uptake experiments (Chapter 3.3.3). Therefore, kaolinite and goethite were chosen as model minerals: their sorption characteristics are known to strongly influence Se bioavailability, they have been studied extensively and their properties are well known, as opposed to oganic matter.

4.1.2 Properties of kaolinite

Because surface functional groups influence both chemical and electrostatic properties, which are defining for the kind of sorption processes that can occur, sorption materials are classified according to these functional surface groups [Davis & Kent, 1990]. Kaolinite, one of the minerals chosen for this study, belongs to the subgroup of aluminosilicates, which do not posses permanent structural charge, while goethite belongs to the hydrous oxide-bearing minerals. Both minerals have in common that their sorption properties are determined by surface hydroxyl groups [Davis & Kent, 1990].

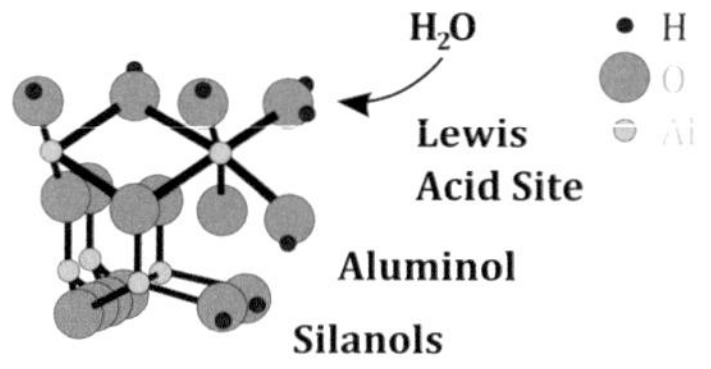

Figure 4.1: Types of surface hydroxyl groups on kaolinite: OH groups at the basal plane, Lewis acid sites (at which H_2O is adsorbed), aluminol and silanol groups [Sposito, 1984, Davis & Kent, 1990]

Kaolinite ($Al_2Si_2O_5(OH)_4$) is a phyllosilicate mineral, in which each $Al(III)O_6$-containing octahedral layer is linked to an $Si(IV)O_4$-containing tetrahedral layer. It is often found in soils in warm, moist climates as a result of deep weathering [Singh & Gilkes, 1998]. Kaolinite is the clay mineral of choice when studying the anion adsorption and desorption properties of clay minerals because, ideally, it has no layer charge or inter-layer cations to enable inter-layer cation exchange [Lagaly &

Köster, 1993]. As a consequence of its well-packed structure, kaolinite particles are not easily broken down and the kaolinite layers are not easily separated. Therefore, Se anion adsorption onto kaolinite occurs mainly at the outer edges and surfaces of the structure [Bar-Yosef & Meek, 1987, Miranda-Trevino & Coles, 2003]. Because kaolinite is known to have a much greater proportion of its total surface area as broken edges, of all clay minerals commonly found in soils, it would be expected to adsorb more Se than any other [Dhillon & Dhillon, 2009].

The degree of ionic substitution within kaolinite layers is less than 0.01 ions per unit cell, resulting in low permanent charge [Sposito, 1984]. As shown in Figure 4.1, kaolinite has three different types of surface hydroxyl groups: aluminol (=Al-OH), silanol (−Si OH) and adsorbed water at Lewis acid sites [Sposito, 1984, Davis & Kent, 1990]. This has direct implications for the pH_{PZC}. Often, published pH_{PZC} values for kaolinite are around 3 or 4 [Schroth & Sposito, 1997], [Appel et al., 2003] [Miranda-Trevino & Coles, 2003]. However, there is little concensus on this, as a wide array of values for kaolinite's pH_{PZC} have been published (pH 2 - 6), as well as multiple pH_{PZC}s or the possibility of none at all [Schroth & Sposito, 1997, Yukselen & Kaya, 2002]. This is attributed to the fact that the lower pH_{PZC} value, i.e. pH 4.2, corresponds to the surface groups [Hur & Schlautmann, 2003], while the higher pH_{PZC} value of 6 - 7.2 corresponds to the amphoteric edge groups [Hur & Schlautmann, 2003, Tombácz & Szekeres, 2006]. This means that above pH 7, kaolinite surface sites are negatively charged and anion adsorption is likely to occur as ligand exchange [Tombácz & Szekeres, 2006]. Below pH 6, however, adsorption is expected to be strongly dependent on the prevalence of surface over edge sites if it occurs as ligand exchange. In any case, coulombic attraction makes physisorption below pH 7 possible [Tombácz & Szekeres, 2006].

4.1.3 Properties of goethite

Goethite (α-FeOOH) is a ferric oxy-hydroxyl mineral belonging to the hydroxide minerals. It is another mineral frequently created in soils by weathering [Lagaly & Köster, 1993] and is one of the most common and reactive crystalline iron hydroxide phases found in soils and sediments [Ghose et al., 2010].

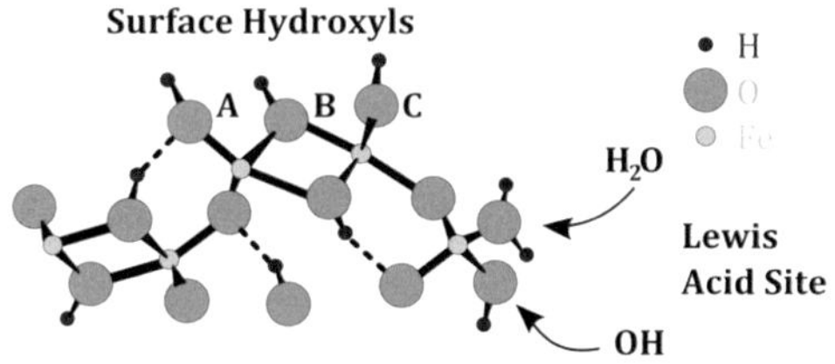

Figure 4.2: Types of surface hydroxyl groups on goethite: hydroxyl groups coordinated singly (A), doubly (B) and triply (C) with Fe(III) ions and one Lewis acid site [Sposito, 1984, Davis & Kent, 1990]

Because of goethite's importance as a ubiquitous environmental substrate, it is commonly studied to investigate interfacial processes such as sorption, dissolution, aggregation, and precipitation [Ghose et al., 2010]. Its adsorptive surface properties are described by four types of surface hydroxyl groups (Figure 4.2) [Davis & Kent, 1990]. One of these is a designated Lewis acid site, at which a water molecule is chemisorbed directly onto an Fe(III) atom. Reactivity of the other three sites is dependent on whether the O-atom in the FeOH group is coordinated with 1, 2 or 3 adjacent Fe(III) ions (Figure 4.2) [Davis & Kent, 1990]. Of these four sites, only the hydroxyl group with the once-coordinated O-atom is considered to be a basic site (proton acceptor) [Sposito, 1984].

Values for pH_{IEP} and pH_{PPZC} range from 7.3 to 9.4 [Parks, 1990, Kosmulski, 2002]. Therefore, the three influences on the formation of Se sorption complexes with ferric hydrates – pH, ionic strength and surface coverage – can be summarized as follows [Fukushi & Sverjenski, 2007]:

1. in general, adsorption declines with higher pH values beginning between pH 5 and 8, coinciding with the presence of the deprotonated selenate molecule in the solution;
2. lowering the background ionic strength leads to a shift of adsorption stability toward higher pH values;
3. raising the selenate-to-background-ion-ratio leads to a higher prevalence of the outer-sphere complex.

Depending on surface conditions, pH-range of stability for a certain surface complexation changes. For example, high pH_{PZNPC} and high BET goethite exhibits the highest stability for inner-sphere complexation for low pH values, while the outer-sphere complexation becomes more stable with neutral pH values. In contrast, low pH_{PZNPC} and low BET lead to both forms of complexation being stable at the same pH value. The importance of the outer-sphere species increases with increasing surface coverage, which means that, for equal background ionic strength and equal selenate concentration, less available goethite surface leads to a greater share of more outer-sphere complexation.

4.2 Processes at the mineral/water interface

Interaction of anions in aqueous solution with a mineral surface can be divided into three main categories: (ad)sorption, co-precipitation and surface precipitation processes, the latter two of which are determined by solubility constants [Fernández-Martínez & Charlet, 2009]. Adsorption, in general, is defined as

> *"the process through which a chemical substance accumulates at the common boundary layer of two contiguous phases." [Sposito, 2004, Fernández-Martínez & Charlet, 2009].*

There are two main types of modelling adsorption-desorption equilibria at mineral surfaces [Davis & Kent, 1990, Goldberg et al., 2007]: (1) empirical partitioning relationships, such as the Langmuir or Freundlich isotherm, which are used to describe natural systems; (2) surface complexation models based on ion association used to gain thermodynamic understanding of coordinative properties. The difficulty with either description type is that they each require the state of equilibrium to be applicable [Davis & Kent, 1990].

4.2.1 The Langmuir and Freundlich isotherms

The Langmuir isotherm (Figure 4.3a) was originally developed to describe monomolecular chemisorption of gasses onto mineral surfaces [Langmuir, 1918] but is now also used to describe adsorption processes in aqueous solution (Equation 4.1) [Davis & Kent, 1990].

$$\equiv X + A_{aq} \Leftrightarrow \equiv XA \tag{4.1}$$

with

$\equiv X$	as the adsorptive surface site
A_{aq}	as the adsorbing solute in aqueous solution
$\equiv XA$	as the adsorbed solute species

The following equation (Equation 4.2), known as the Langmuir isotherm (Figure 4.3a), is based on the assumptions that (1) all surface sites have the same affinity for the solute, (2) X is the only adsorbing solute, (3) there is no lateral solute interaction and (4) the surface site density is finite [Davis & Kent, 1990, Dörfler, 2002].

$$\Gamma_A = \Gamma_{max} \frac{K_L\, c(A_{aq})}{1 + K_L\, c(A_{aq})} \tag{4.2}$$

for

$$K_L = \frac{\Gamma_A}{c(A_{aq})\, \Gamma_X} \tag{4.3}$$

with

Γ_A	as the adsorption density of the solute [g/kg]
Γ_{max}	as the maximum surface site density [g/kg]
Γ_X	as the surface density of uncomplexed sites [g/kg]
K_L	as the conditional Langmuir equilibrium constant [-]
$c(A_{aq})$	as the concentration of solute in the solution [g/L]

Although the Langmuir isotherm is a mathmatical formula based on equilibrium phase observations, a good fit of experimental data to the Langmuir isotherm provides no insight on the actual mechanisms on the mineral surface [Davis & Kent, 1990]. Special cases of precipitation, for instance, have also been shown to confirm to the Langmuir isotherm [Davis & Kent, 1990].

Not only does the Langmuir isotherm not exclusively describe adsorption processes but, furthermore, not all sorption processes are described by the Langmuir isotherm. Inorganic anions, for example, frequently are not described by the Langmuir isotherm as the requirement of a monolayer adsorption is not met [Davis & Kent, 1990, Dörfler, 2002].

Therefore, the data is either fitted to a multiple-site Langmuir expression or to a generalized exponential term such as the Freundlich isotherm (Figure 4.3b, Equation 4.4) [Davis & Kent, 1990, Dada et al., 2012].

$$\Gamma_A = K_F \, c(A_{aq})^n \tag{4.4}$$

with

Γ_A	as the adsorption density of the solute [g/kg]
K_F	as the conditional Freundlich equilibrium constant [-]
$c(A_{aq})$	as the concentration of solute in the solution [g/L]
n	as the conditional Freundlich exponent [-]

Two other frequently used isotherms are the Temkin isotherm, which takes into account adsorbent-adsorbate interactions and the Dubinin-Radushkevich isotherm, which is often successful for high solute activities [Dada et al., 2012]. Although these empirical approaches are simple and can easily be fitted to many data sets, there are drawbacks as well. For instance, the statistically-based empirical approach does not typically lead to a general mechanistic understanding [Goldberg et al., 2007]. Furthermore, calculated values for constants such as K_F or K_L are only valid for the specific experimental conditions under which the data was obtained, making general conclusions sketchy at best.

4.2.2 The Electrical Double Layer Model (EDL)

In contrast to empirical adsorption fittings, surface complexation models such as the Electrical Double Layer Model (EDL) have a more mechanistic approach and are, therefore, generally more robust and apply to a greater variety of geochemical conditions. Using mass action laws, chemical speciation and competitive adsorption can be taken into account [Davis & Kent, 1990, Goldberg et al., 2007].

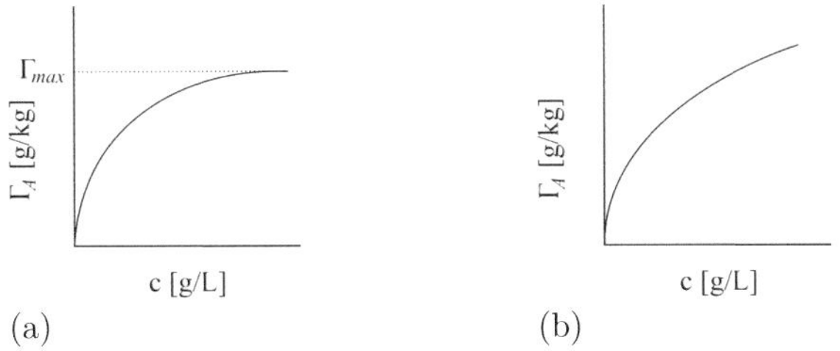

Figure 4.3: Empirical modelling of adsorption-desorption equilibria at mineral surfaces [Goldberg et al., 2007]
a: Langmuir isotherm
b: Freundlich isotherm

When submerged in aqueous solution, any mineral surface becomes an interface between bulk mineral and bulk water properties [Parks, 1990]. This leads to hydroxylation, in which water dissociates and charged, dangling bonds of the mineral's surface form chemical bonds with the resulting hydroxyl groups (OH^-) or protons (H^+). Layered around these hydroxyl or proton bonds to the mineral surface, clusters of water molecules form by hydrogen-bonding, a phenomenon called hydration [Parks, 1990]. While the first monolayer of water is chemically bound to the mineral, the following monolayers are bound increasingly loosely, until water molecules reach bulk water properties.

There are three types of mineral surface charge [Davis & Kent, 1990]:

1. permanent structural charge σ_s, usually as a result of isomorphic substitution (i.e., Al^{3+} instead of Si^{4+});

2. coordinative surface charge σ_o as a result of potential-determining ions coordinated with surface functional groups;

3. dissociated surface charge σ_d, which is provided by counter-ions and preserves electroneutrality when the net particle charge σ_p is not zero, as shown in Equation 4.5.

$$\sigma_p + \sigma_d = 0 \tag{4.5}$$

and

$$\sigma_p = \sigma_s + \sigma_o \tag{4.6}$$

with

σ_p	as the net particle charge [C/m^2]
σ_d	as the dissociated surface charge [C/m^2]
σ_s	as the permanent structural charge [C/m^2]
σ_o	as the coordinative surface charge [C/m^2]

The surface layer and disassociated layer are referred to as the EDL. Separation of charges in the EDL leads to an electrical potential difference across the particle-water interface [Davis & Kent, 1990]. In the Guoy-Chapman theory (Figure 4.4), this charge potential decays exponentially with the distance from the surface, as the diffuse layer of counter-ions are arranged according to their polarity in coulombic interaction around the charged surface, in effect shielding the surface charge from the surrounding bulk solution [Parks, 1990, Davis & Kent, 1990]. This theory was later refined in the Stern-Grahame theory (Figure 4.4), in which a layer of specific adsorption – or compact layer ("Stern" layer) – was added to account for finite ion size and counter-ion interactions directly at the surface [Davis & Kent, 1990].

In EDL models, all chemical adsorption is accounted for at the x_o plane (Figure 4.5a), which can be equated with chemical bonding directly at the particle surface. However, different types of sorption result in different distances from the particle surface. The triple layer model (TLM) includes a second plane x_β (Figure 4.5b), at which outer-sphere complexes or ions with a greater crystallographic radius are accounted for, as opposed to the inner-sphere complexes and ions of smaller crystallographic radius, which are located at the x_o plane [Goldberg et al., 2007].

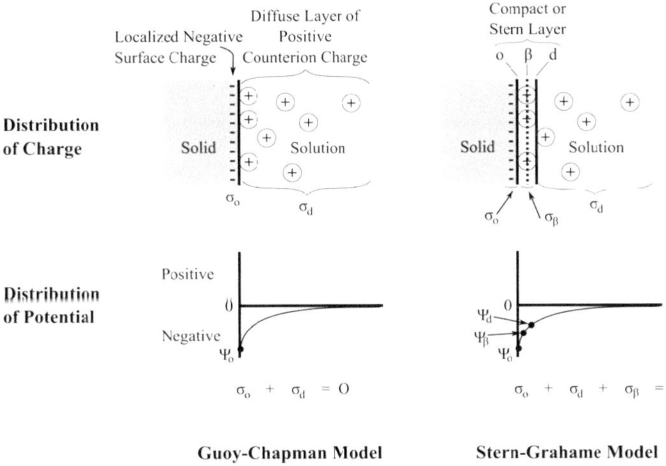

Figure 4.4: Schematic drawing of the electrical double layer of the classical Guoy-Chapman and Stern-Grahame models under the assumption that $\sigma_s = 0$ [Davis & Kent, 1990]

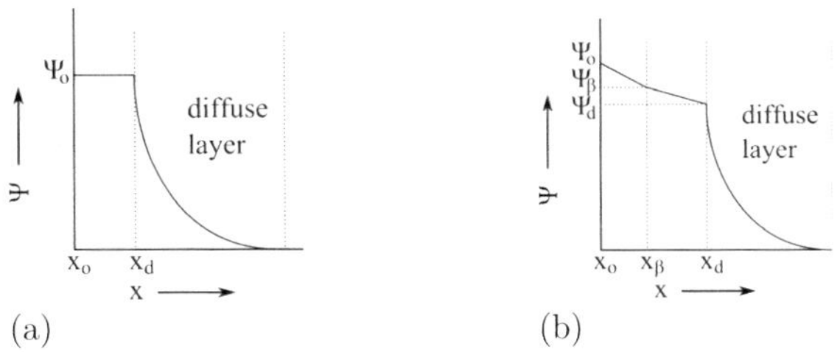

Figure 4.5: Comparison between two surface complexation models [Davis & Kent, 1990, Goldberg et al., 2007]
a: the Electrical Double Layer model (EDL)
b: the Triple Layer Model (TLM)

4.2.3 Points of Zero Charge

Due to a charged surface or dangling bonds, when a solid is submerged in a solution, the initial solution pH value (pH_i) is likely to change unless it equals the so-called critical pH value (pH_{crit}). According to the Stern-Grahame theory, protons and hydroxyl groups act as the coordinative surface charge (Equation 4.7).

$$\sigma_H = \Gamma_{H^+} - \Gamma_{OH^-} \tag{4.7}$$

with

σ_H as proton surface charge [C/m^2]

Γ_{H^+} as positive charge from adsorbed protons [C/m^2]

Γ_{OH^-} as negative charge from adsorbed hydroxyl groups [C/m^2]

If pH_i is greater than pH_{crit}, the solid is an H^+ donor (or OH^- acceptor); if pH_i is smaller than pH_{crit}, the solid is an OH^- donor (or H^+ acceptor). If the exchanged ions include only OH^- or H^+, this pH_{crit} is also known as the Point of Zero Net Proton Charge (PZNPC or pH_{PZNPC}) and describes the pH value at which $\sigma_H = 0$ [Parks, 1990, Davis & Kent, 1990, Sposito, 1998].

Under natural conditions, however, the solution surrounding the solid includes ions other than H^+ or OH^- which shield the surface charge from the surrounding bulk solution. Therefore, protons and hydroxyl groups are only a part of the coordinative surface charge σ_o (Equation 4.8) [Parks, 1990, Davis & Kent, 1990].

$$\sigma_o = \sigma_H + \sigma_{CC} \tag{4.8}$$

with

σ_o as the coordinative surface charge $[C/m^2]$

σ_H as the proton surface charge $[C/m^2]$

σ_{CC} as the coordinative complex surface charge $[C/m^2]$

This means that in an environment of strongly adsorbing ions, the pH_{PZNPC} may shift considerably. This made the definition of the pristine point of zero charge (pH_{PPZC}) necessary, which is defined for the condition $\sigma_o = 0$ [Davis & Kent, 1990, Sposito, 1998].

If a particle and its diffuse layer moves through an electric field or fluid flows past it, an immobile hydrodynamic boundary layer forms. When counter-ions of the diffuse layer around the particle are not within this boundary layer, they are stripped away from the particle, resulting in a charge separation. The particle is then left with a net charge (electrokinetic net charge σ_{ek}), leading to the formation of a potential gradient (zeta potential ζ or Ψ_z) in movement direction [Parks, 1990].

The pH value at which a colloidal particle is electro-kinetically uncharged ($\sigma_{ek} = 0$) and at which, therefore, the potential gradient is also zero ($\Psi_z = 0$) is defined as the iso-electrical point (IEP or pH_{IEP}). At pH_{IEP} or at any pH in the presence of high ionic strength, the diffuse layer collapses ($\sigma_d = 0$) and agglomeration or flocculation of particles can occur. With pH_{IEP} defined as $\sigma_e k = 0$ (also $\sigma_d = 0$), pH_{PPZC} defined as $\sigma_o = 0$ and pH_{PZNPC} defined as $\sigma_H = 0$, all these critical pH values are analogous in concept, but quantitatively not identical [Parks, 1990]. Furthermore, either is difficult to determine experimentally [Sposito, 1998].

4.2.4 Selenium adsorption & complexation

When a mineral surface is surrounded by aqueous solution, (ad)sorption is defined as

> *"chemical species attachment to a mineral surface" [Hochella & White, 1990].*

Depending on the nature of this attachment, adsorption can be categorized into two basic types [Manceau & Charlet, 1994], outer-sphere and inner-sphere complexes, resulting as a consequence of the two concepts of zero charge, the pH_{IEP} and pH_{PZNPC}.

1. *no presence of strongly adsorbing ions* $\rightarrow pH_{PZNPC} = pH_{IEP}$. These two points of zero charge can be equal only if strong adsorption of electrolyte ions by a charged colloid does not occur [Sposito, 1998]. In this case, anions adsorb when $pH < pH_{PZNPC}$ and Ψ_z is positive (cations adsorb when $pH > pH_{PZNPC}$). There is minimal adsorption at $pH = pH_{PZNPC}$ when the surface is uncharged because adsorption occurs predominantly through coulombic attraction and part of the diffuse layer remains between the ion and particle surface [Parks, 1990]. This leads to outer-sphere surface complexation, which describes the process of non-specific adsorption (physical adsorption), in which a water molecule is retained between the adsorbing ligand and the surface site. The water molecule is thus part of the adsorption complex, which means that the bonding of the ligand to the surface is of electrostatic nature rather than co-valent or ionic. Characterized by hydrogen bridge bonds or van-der-Vaals interactions, this type of complex is, therefore, less stable than the inner-sphere complex [Parks, 1990].

2. *in the presence of strongly adsorbing ions* $\rightarrow pH_{PZNPC} \neq pH_{IEP}$. In the presence of strongly adsorbing ions, adsorption is insensitive to surface charge and allows super-equivalence, which describes adsorption of greater density than the electrical equivalent of the surface charge density [Parks, 1990]. Ion adsorption on surfaces of like charge is electrostatically repelled, which is why the strength of the bond must exceed this electrostatic repulsion [Parks, 1990]. This leads to inner-sphere surface complexation, which describes the process of specific adsorption (chemical adsorption), in which an aqueous ligand, i.e. an anion, exchanges for a surface group, i.e. hydroxyl group. First, the surface hydroxyl group is protonated by a proton of the positively charged aqueous hull around the adsorbing molecule. Then, the protonated hydroxyl group is split off as a water molecule and replaced by the adsorbing anion [Parks, 1990]. The chemical bond between the surface site and the ligand can be co-valent or ionic or in-between, which explains the greater stability of this complex [Fernández-Martínez & Charlet, 2009].

Both selenite and selenate are known to adsorb onto ferric hydroxides [Dhillon & Dhillon, 2009]. However, affinity of each anion toward these solids differs. For a long time, anions were simply classified according to their affinity to oxide surfaces, which was believed to determine the nature of the chemical attachment [Manceau & Charlet, 1994]. High-affinity anions including selenite and phosphate were thought to be bound strongly to the sorbent by inner-sphere surface complexes. Low-affinity anions like nitrate, selenate and sulfate were thought to be bound weakly to the sorbent by an outer-sphere complex.

This convenient explanation holds true for selenite, which, in general, is adsorbed to oxy-hydroxides as shown in Figure 4.6a by a bidentate inner-sphere surface complex [Manceau & Charlet, 1994, Su & Suarez,

2000, Fernández-Martínez & Charlet, 2009]. However, while many authors have confirmed that selenate has less affinity to ferric oxy-hydroxides than selenite [Davis & Kent, 1990, Hayes et al., 1987], through the use of modern spectroscopic techniques, such as EXAFs, it was discovered in 1994 that the traditionally accepted macroscopic link between ionic dependence of adsorption and outer-sphere complexation did not always apply. While it is true that outer-sphere complexes are strongly dependent on ionic strength and are much more sensitive to competing ions which can replace the outer-sphere-adsorbed ligand, the inverse conclusion was shown to be false. Sensitivity to ionic strength is not an adequate macroscopic criterion with which outer-sphere complexation can be determined [Manceau & Charlet, 1994].

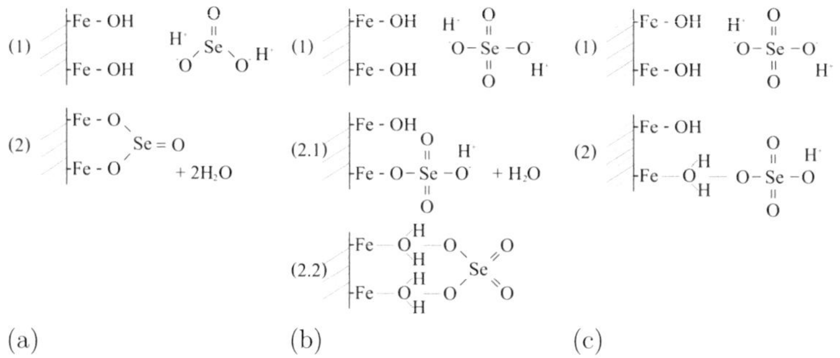

(a) (b) (c)

Figure 4.6: Adsorption of Se oxy-anions in 2 steps; step 1 shows the free anion in solution, step 2 shows resulting surface complexation [Fukushi & Sverjenski, 2007]
a: Specific adsorption of selenite as ligand exchange forming a bidentate-mononuclear inner-sphere complex
b: Complexation of selenate, either as specific adsorption as a monodentate-mononuclear inner-sphere complex (step 2.1) or as non-specific adsorption as a bidentate-binuclear outer-sphere complex (step 2.2)
c: Selenate adsorption onto amorphous ferric oxide as bidentate-mononuclear outer-sphere complex

In an attempt to reconcile conflicting data on selenate complexation on goethite from ATR-FTIR and EXAFS studies [Hayes et al., 1987, Manceau & Charlet, 1994, Su & Suarez, 2000], the extended triple layer model (ETLM) took into account the substantial electrostatic work associated with the desorption of water dipoles during ligand exchange reactions [Fukushi & Sverjenski, 2007]. As shown in Figure 4.6b, selenate shows two main adsorption mechanisms onto goethite, a monodentate-mononuclear inner-sphere and a bidentate-binuclear outer-sphere (or H-bonded) complex, the relative proportions of which are functions primarily of pH and ionic strength [Fukushi & Sverjenski, 2007]. However, as shown in Figure 4.6c, selenate adsorption onto hydrous ferric oxide, an amorphous goethite relative, is characterized by a bidentate mononuclear outer-sphere complex (also H-bonded).

4.2.5 Se adsorption competition by nutrient anions

For the purpose of this study, the following main nutrient anions are of great interest: sulfate (SO_4^{2-}), phosphate (PO_4^{3-}) and nitrate (NO_3^-). In natural soil solutions, each of these anions is present in much higher concentration than either selenite or selenate. While Se in soils is roughly 0.1 - 2 mg/kg [Fernández-Martínez & Charlet, 2009], concentrations of S as sulfate are generally 2 - 20 mg/kg [Horneck et al., 2011], extractable concentrations of P are usually 10 - 100 mg/kg [Horneck et al., 2011] and plant-available concentrations of N are generally around 2 - 10 mg/kg as NH_4^+ and 10 - 30 mg/kg as NO_3^- [Marx et al., 1999]. Therefore, competition by these ions is an important aspect of Se sorption processes and influences the Se cycle.

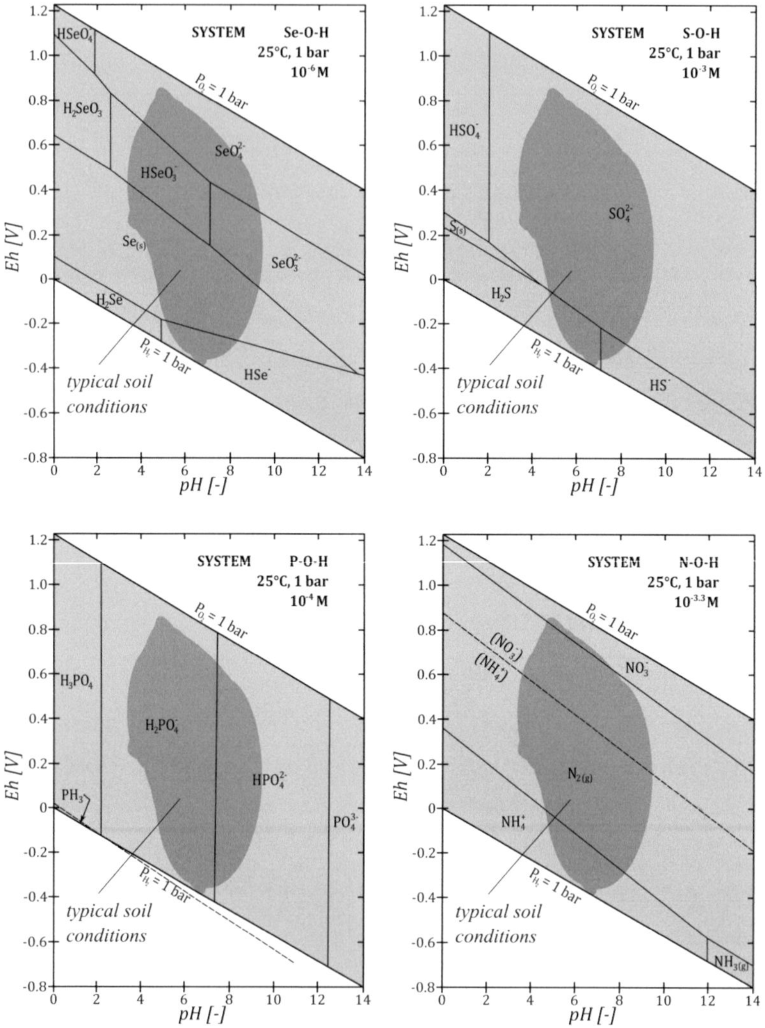

Figure 4.7: Eh-pH diagrams for thermodynamic data of Se (10^{-6} mol/L), S (10^{-3} mol/L), P (10^{-4} mol/L) and N ($10^{-3.3}$ mol/L) within the stability field of water at 25 °C at 1 bar [Brookins, 1988] with typical soil conditions added according to [Neal, 1995]

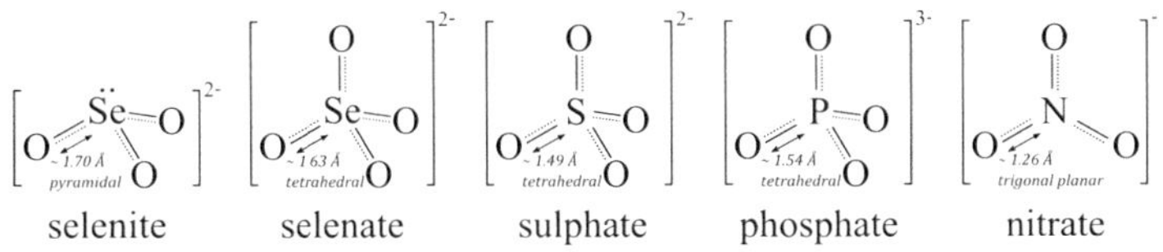

Figure 4.8: 3D structure and X-O bond lengths of Se oxy-anions selenite and selenate as well as the three main nutrient oxy-anions phosphate, sulfate and nitrate [Kálmán, 1971, Paul & Pryor, 1972, Séby et al., 2001, Mortimer, 2010]

In order to determine how these anions affect selenite and selenate adsorption, redox behaviour and pH-dependent speciation information is necessary. While Se occurs both as selenite and selenate in typical oxidized soil conditions at pH 4 - 10, S is mainly found as sulfate, P is protonated either as di-hydrogen or hydrogen phosphate and N is found as nitrate or ammonium (Figure 4.7). As shown in Figure 4.8, these anions also all exhibit distinct structural properties. Selenite is pyramidal with its two free electron pairs forming a semi-tetrahedral structure, while selenate, sulfate and phosphate are all tetrahedral in structure [Kálmán, 1971, Séby et al., 2001, Mortimer, 2010]. Nitrate, on the other hand, is a trigonal planar molecule [Paul & Pryor, 1972]. Electronegativity differences between the central atom and the O-atom in these anions, which, according to Pauling [Mortimer, 2010] account for polarizability of a molecule are the highest in phosphate (P-O bond, $EN_{diff} = 1.2$) and the lowest in nitrate (N-O bond, $EN_{diff} = 0.2$). sulfate and selenate are alike (S-O and Se-O bond, $EN_{diff} = 0.8$), but selenite has an additional free electron pair (Figure 4.8), which increases its molecular polarization [Mortimer, 2010].

Structural and electrochemical similarity between sulfate and selenate is expected to lead to similar sorption behaviour [Fernández-Martínez & Charlet, 2009]. Competition effects between selenite and phosphate have been observed, leading to the conclusion that sorption mechanisms of ligand exchange are similar between these two ions [Neal, 1995]. Due to greater differences in structure and size between nitrate and Se oxy-anions, nitrate is not expected to show the same competition effects as phosphate or sulfate. Nitrate is, therefore, believed to bind as an outer-sphere complex [Neal, 1995]. In conclusion, the affinity sequence for selected anion adsorption of nutrient oxy-anions and Se oxy-anions onto soils is considered to be the following:

phosphate > selenite > > sulfate > selenate > nitrate [Neal, 1995]

4.3 Material characterization of kaolinite and goethite

Characterization of sorption processes onto the surface of a mineral requires a certain degree of purity in order to exclude interfering sorption phenomena and/or leaching of the chosen element from the substrate. Therefore, the materials used for the following experiments were analyzed for their purity. Table 4.1 summarizes the obtained industrial products, which were assessed for their quality with standard analytical procedures.

Table 4.1: Industrial products tested for quality as sorption material

sample name	**main mineral**	**company**
Kaolin AKW KN 83	kaolinite	Amberger Kaolinwerke
Kaolin KBS Opal Alpha	kaolinite	Kruse Bassermann Specialties
Kaolin KBS Burgess Thermo-Glace HB	kaolinite	Burgess Pigment
FeOOH Bayoxide E 216	ferric oxy-hydroxide	Bayferrox
FeOOH Bayoxide 920 Z	ferric oxy-hydroxide	Bayferrox
Dors 9S 1-0.6	quartz	Dorsillit

A total of three industrial kaolin samples was ordered from two different companies with the request for good sorption properties and high kaolinite purity. Two different industrial goethite samples and a quartz sample were also analyzed for their purity and sorption properties. Since kaolin is a natural material containing varying quantities of kaolinite and the goethite ordered was an industrial product of unknown refinery, quality assessment was a prerequisite for all the following experiments. Of the three kaolinite and two goethite products, one product of each mineral with the greatest purity was selected.

4.3.1 Sample preparation

All materials were analyzed for their chemical components by means of full acid digestion in Teflon beakers, with 0.1 g sample incubated with 2 mL of 65 % HNO_3 (VWR 20429.320 p.a. sub-boiled) at 200 °C for 30 min before being concentrated in multiple steps with 40 % HF (Merck 1.00335.1000, suprapure) and $HClO_4$ (VWR 1.00517.1000, suprapure). Digested solutions were then measured with ICP-MS (Chapter 3.4.4). Furthermore, fused beads of the samples were analyzed with Wavelength dispersive X-ray spectroscopy (WDX). Sample preparation of the fused beads follows a standardized method [DIN EN ISO 12677:2013], using 0.5 g of freshly ground sample dried at 40 °C with 5 g of 65.5 % dilithiumtetraborate and 34.5 % lithiummetaborate fluxing agent (111802 Spectromelt A12 from Merck).

In-depth kaolinite analysis was conducted with thermogravimetry coupled with differential scanning calorimetry (TG-DSC) in cooperation with members of the Competence Centre for Material Moisture (CMM, KIT) using an STA 449C Jupiter (NETSCH). After a reference measurement run-through with the empty Pt-crucible, 100 mg of sample powder was tared, measured isothermically at a temperature of

35 °C for room-temperature water evaporation and then analyzed for its temperature profile from 35 - 1100 °C. To verify that the weight loss attributed to kaolinite restructuralization was indeed due to crystal water loss, the resulting gas flow during the TG-DSC measurement was analyzed with an MS (Infrared mass spectrometer, 2MS 403C Aëolas). This was used to identify ion counts for masses M = 18 (H_2O), M = 44 (CO_2), M = 16 (O) and M = 64 (SO_2) in 50 mL of sample gas combined with 20 mL N_2 protective gas.

Further evidence of clay mineral content in the kaolin samples was obtained by measuring texture samples [Tributh, 1976, Lagaly & Köster, 1993], prepared as follows: 50 mg of sample powder was filled into a 15-mL glass test tube, submerged in 10 mL of diluted NH_4OH (10 mL NH_4OH per 1 L double-deionized water), shaken and put into an ultrasound bath for 15 min. After shaking, the sample was left to sediment for 1 h. With the non-clay particles sedimented and the clay particles held in submersion by NH_4OH, 3 1-mL samples of the supernatent were pipetted onto a glass object slide (diameter: 2 cm). The first of these three samples was air-dried for 24 h as a reference, the second was treated with ethylene glycol for 24 h to test for expanding layers of non-kaolinite clays and the third was burnt at 550 °C for 3 h to test for the kaolinite-typical collapse of the reference kaolinite peak after structural change due to water loss.

For the final step in the kaolin material characterization, an extensive X-ray diffraction scan (XRD) was performed at CMM using a Siemens D5000 X-ray Powder Diffraction System (angular range: 3 - 100 °, step-size: 0.01 °, time step 0.1 s, anode material: Cu) Using DIFFRAC.SUITE EVA (Bruker) in combination with AutoQuan (BGMN), the mineral phases found in any of the three different kaolin powders (kaolinite, muscovite, orthoclase, christobalite, quartz, tridymite, calcite, magnetite, anatase and rutile) were used for the computation of a theoretical XRD diffractogram, which can be subtracted from the measured diffractogram to account for unidentified phases.

Goethite characterization was achieved by X-ray diffraction (Siemens D500 X-ray Powder Diffraction System, angular range: 2 - 22 °, stepsize: 0.01 °, time step 1 s, blinds: 1.54, anode material: Cu) for both industrial FeOOH samples. After chosing the purest kaolinite and goethite, their surface areas were calculated using a N_2-Brunauer-Emmett-Teller scan (BET) performed by Peter Weidler, IFG (KIT), using a Quantachrome Autosorb 1-MP instrument. Additionally, scanning electron microscopy images (SEM) were taken in cooperation with Frank Zibat, LEM, (KIT).

4.3.2 Material characterization results for kaolinite and goethite

Figure 4.9 shows component percentages as oxide-equivalents corrected by the loss of mass by ignition LOI_{1050}, with the exception of quartzsand, for which loss by ignition was not analyzed since it was assumed to be zero. For better comparison, minor component percentages (< 5 %) are shown separately.

The Dorsillit quartz sample showed a high degree of purity (99.1 %) and was used for experiments with open rice cultures (Chapter 3.3.1). Kaolin samples KBS BTG HB and AKW KN 83 were similar in purity according to WDX- and ICP-MS analysis (Figure 4.9), while the kaolin sample KBS 0 alpha was clearly less pure than the others due to a high K content. In the goethite samples, a large quantity of quarz in the 920 Z sample significantly lowered the goethite percentage of the substrate after loss by ignition.

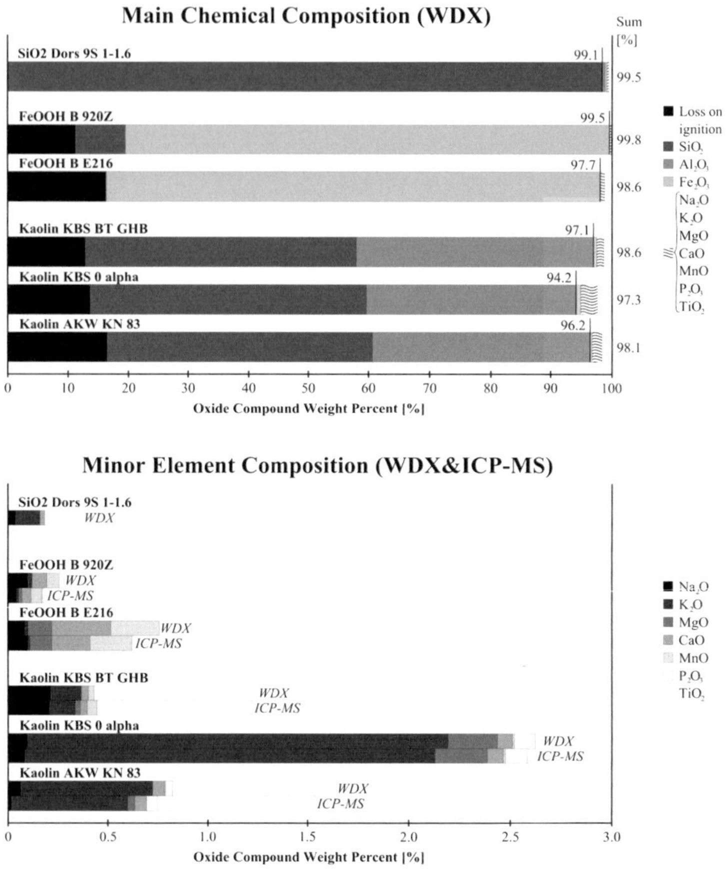

Figure 4.9: Main chemical composition (top) and minor element composition (bottom) of sorption materials in oxide-equivalents as measured with WDX and ICP-MS after full HF-digestion

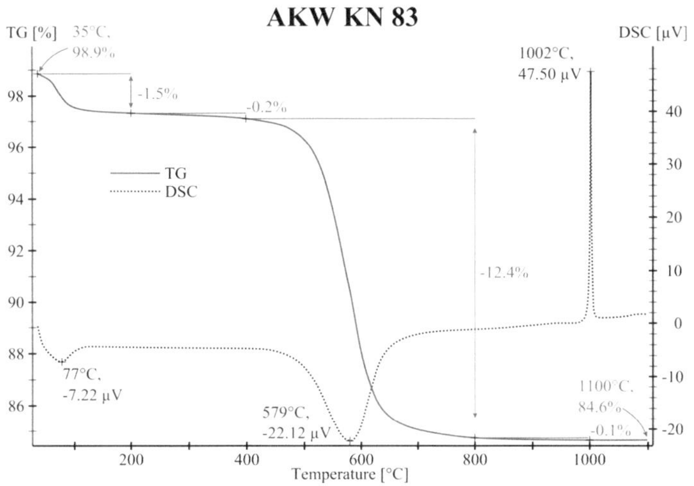

Figure 4.10: TG-DSC curve of kaolin powder AKW KN 83 measured at CMM (100 % kaolinite yields a weight loss at 550 °C of 13.95 %, therefore 12.4 % weightloss indicate a Kaolinite purity of 89 % for this sample

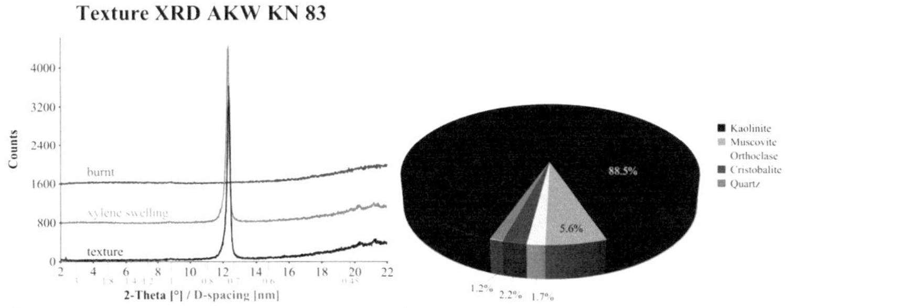

(a) (b)

Figure 4.11: Characterization of kaolin sample AKW KN 83
a: XRD analysis of texture samples untreated, burnt and swelled
b: AutoQuan-calculated mineral composition, calculated total: 99.2 % ±0.5

Figure 4.10 shows the resulting TG-DSC curve (analyzed with NETSCH Proteus-Thermal-Analysis Version 5.1.0 Software) for kaolin powder AKW KN 83.

Using the weight loss at 550 °C to calculate kaolinite purity (Figure 4.10), the AKW KN 83 sample was chosen for its purity of 89 % kaolinite, compared to 78 % for KBS 0 alpha and 86 % for KBS BTG HB for all future sorption experiments. Verification of the nature of the evaporated gases (H_2O, O_2, SO_4 and CO_2) through the coupled MS showed that for AKW KN 83, gas loss was mainly water and a small amount of CO_2, while the other two samples had measurable amounts of CO_2 and SO_2.

Analysis of the AKW KN 83 texture samples with XRD (Figure 4.11a) revealed no other clay minerals beside kaolinite, as xylene swelling did not change the diffractogram and the kaolinite-typical peak completely collapsed in the burnt samples. This was also the case for KBS BTG HB, but the sample for KBS 0 alpha showed peaks in the xylene swelling sample, suggesting the presence of 2-layer clay minerals. Figure 4.11b shows the calculated mineral composition of AKW KN 83 using AutoQuan calculations of the XRD. The kaolinite purity of this kaolin sample is considered to be the mean value of TG-DSC analysis and AutoQuan calculation: 88.7 %.

Goethite peaks in the XRD diffractogram (Figure 4.12) are similarly positioned in both samples, indicating that both are primarily comprised of α-FeOOH, also known as goethite. However, sample E216 appears to contain a larger amorphous share, since the peaks are not as clearly defined as those of sample 920Z. Due to a large likelihood of amorphous phases in natural soils and the higher share of goethite in the sample, the following sorption studies were conducted with the more amorphous and purer E216.

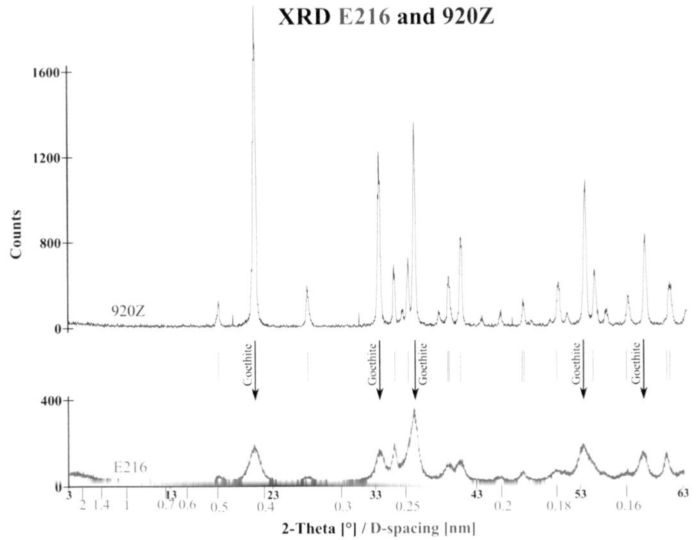

Figure 4.12: XRD analysis of FeOOH samples: the more amorphous Bayferrox E216 and highly crystalline Bayferrox 920Z; main goethite-specific peaks are indicated with lablled arrows, position of minor goethite peaks indicated with lines

The BET scans of both goethite samples yielded a surface of 126.4 m^2 $\pm$1.8 for E216 and 10.0 m^2 $\pm$0.5 for 920Z, while kaolinite was found to have a surface of 9.4 m^2/g $\pm$0.1. The SEM images of the mineral surfaces for kaolinite (Figure 4.13a) and goethite (Figure 4.13b) shows typical, well-crystallized surfaces for kaolinite with sheet-like structures, while the large amorphous share and much larger surface area is visible for goethite.

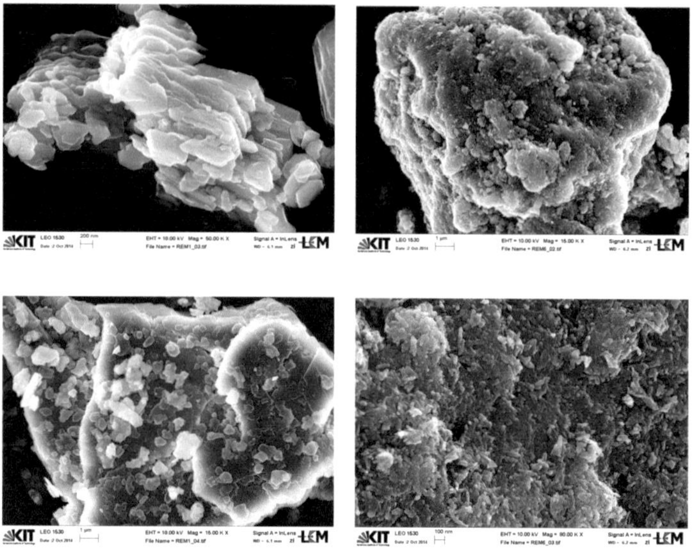

(a) (b)

Figure 4.13: SEM images taken from sorption materials (electrical conductivity provided by thin Pt-coating)
a: kaolinite AKW KN 83, magnification 15 K (above), 50 K (below)
b: goethite E216, magnification 15 K (above), 80 K (below)

4.4 Experimental set-ups

The following sorption studies are divided into 4 main parts:

1. preliminary tests;
2. pure selenite and selenate adsorption onto the substrates kaolinite and goethite;
3. subsequent desorption of previously adsorbed Se from kaolinite and goethite;

4. selenite and selenate adsorption in the presence of the competing anions phosphate, sulfate and nitrate. Based on results from the preliminary tests, experimental set-ups were optimized.

4.4.1 Preliminary tests

In order to predict which sorption concentrations and substrate amounts might yield measurable results in later experiments the following sets of preliminary tests were conducted.

Substrate reaction to double-deionized water

Before sorption experiments were conducted, pH value changes due to mineral surface properties were measured using 0.5 g (Sartorius 1712MP8) of either kaolinite or goethite and 10 mL of double-deionized water in 20-mL patho vessels (Böttger 08-313-1001). The pH values of two double-deionized water blanks, two kaolinite samples and two goethite samples were measured in five replicates with two pH electrodes (WTW SenTix 41, WTW SenTix V), due to the difficulties of measuring the pH value of double-deionized water.

Sorption influence of substrate amount vs. Se concentration

For the sorption experiments, kaolinite and goethite powders were milled for 5 min to a homogeneous powder using an agate mill (Scheibenschwingmühle-TS, Siebtechnik). The first set of preliminary sorption studies was carried out using kaolinite or goethite amounts of 50, 100, 500 and 1000 mg (Sartorius basic BA 1105) weighed into 15-mL centrifuge tubes (VWR 525-0149). Concentrations of 50, 500 or 5000 µg/L Se as selenate were achieved by pipetting 10, 100 or 1000 µL of 500 mg/L

Se stock solution (in an Eppendorf reaction tube: 1 mL of double-deionized water and 1 mL of $Na_2SeO_4 \cdot 10H_2O$, VWR BDH Prolabo 302113L) into 100 mL glass flasks prepared with 3.536 mmol $CaCl_2$ solution (194.6285 mg $CaCl_2$ in 500 mL of double-deionized water) as background ionic strength comparable to the nutrient solution (Chapter 3.3.3). Ten mL of sorption solution was added to each substrate sample and shaken for 24 h at 180 rpm. After sorption tubes were centrifuged at 40,000 rpm for 5 min (Rotofix 32 A), 9.5 mL of sorption solution from kaolinite samples and 9.25 mL from goethite samples was extracted, pipetted into labelled 15-mL centrifuge tubes (VWR 525-0149) and prepared for HG-FIAS measurement (Chapter 3.4.3).

Experimental desorption properties

In this set of preliminary tests, Se losses due to the substrate washing step preceding the desorption experiment needed to be quantified. Furthermore, the ability of an extraction step [Kulp & Pratt, 2004], [Bacon & Davidson, 2007] as desorption agent for easily exchangeable Se was explored.

After sampling sorption solution, Se-sorbed kaolinite and goethite samples were prepared for desorption. A washing step to remove residual sorption solution was conducted by resubmerging the substrate in 9.5 mL of double-deionized water for 1 min. After centrifuging at 40,000 rpm for 5 min (Rotofix 32 A), sorption solution (9.5 mL from kaolinite solutions and 9.25 mL from goethite solutions) was sampled into labelled 15-mL centrifuge tubes (VWR 525-0149). Remaining substrate was then resubmerged in double-distilled water (amounts equal to the sampled sortion solution), manually shaken for 1 min before being centrifuged at at 40,000 rpm for 5 min (Rotofix 32 A). This double-distilled water was sampled into labelled 15-mL centrifuge tubes (VWR

525-0149) as well in order to verify any unintended Se loss during this step. Both sorption samples and wash samples were prepared for HG-FIAS measurement (Chapter 3.4.3).

A desorption solution of 0.1 mol/L K_2HPO_4 [Kulp & Pratt, 2004, Bacon & Davidson, 2007] was prepared as follows: in a 500-mL glass bottle (Schott), 8.7087 g of K_2HPO_4 were dissolved in 500 mL of double-deionized water and then brought to a pH of 8.0 using 100 µL (±30) 37 % HCl (Merck, p.a. 1.00317.2500). Each sample was re-sumberged in 9.5 mL of desorption solution and placed on the shaking table at 180 rpm for 24 h. After centrifuging at 40,000 rpm for 5 min, 9.5 mL of sorption solution was sampled into labelled 15-mL centrifuge tubes (VWR 525-0149) and prepared for HG-FIAS measurement (Chapter 3.4.3).

Experimental up-scaling effects

Keeping the ratio of sorption solution volume per substrate amount constant, the following up-scaling experiments were designed to identify whether sorption results were affected by experimental conditions such as bottle size or if a constant ratio between solution volume and substrate amount yielded the same results on different scales. The experiment was carried out using 1, 5, 10, 50 and 100 g of milled kaolinite or goethite and 10, 50, 100, 500 and 1000 mL of 5000 µg/L Se as selenate ($Na_2SeO_4 \cdot 10H_2O$, VWR BDH Prolabo 302113L) in 3.536 mmol $CaCl_2$ in screw-lid PP bottles, respectively. Samples were placed on a shaking table at 180 rpm for 24 h. After centrifuging at 40,000 rpm for 5 min (Rotofix 32 A), 7.5 mL of sorption solution was sampled into labelled 15-mL centrifuge tubes (VWR 525-0149) and prepared for HG-FIAS measurement (Chapter 3.4.3).

4.4.2 Exp. B1 – pure Se sorption onto kaolinite and goethite

The following batch experiment was designed to determine selenite and selenate adsorption properties of kaolinite and goethite in the presence of ionic strength similar to the nutrient solution used in experiments A3 (Chapter 3.3.3).

(a) (b)

Figure 4.14: Photos of sorption experiment preparations
a: weighing kaolinite for selenite adsorption studies (photo taken on 07.07.2014)
b: distribution of 30 mL of sorption solution of selenate adsorption experiments into goethite, kaolinite and sampling patho-vessels with pH measurement (photo taken on 14.05.2014)

As shown in Figure 4.14a, for every batch of 36 samples, 0.5 g of either milled kaolinite (18 samples) or milled goethite (18 samples) were weighed (Sartorius 1712MP8) into labelled 20-mL patho vessels (Böttger 08-313-1001). To ensure that the sampled initial solution was identical for each Se concentration of the respective kaolinite and goethite sorption batch as well as their sampled solution for analysis, each of the 18 sorption solutions of 30 mL (10 mL for kaolinite sorption + 10 mL for goethite sorption + 10 mL for analysis) was prepared in 50-mL glass beakers. Because the range of Se concentrations in this study was so large (0, 10, 25, 50, 75, 100, 150, 200, 250, 350, 500, 750, 1000, 1500, 2000, 2500, 3500 and 5000 µg/L), two Se stock solutions of 10 mg/L and 1000 mg/L Se had to be used; in 7.735 mmol/L KCl, either as $Na_2SeO_4 \cdot 10H_2O$ (VWR BDH Prolabo 302113L), or as Na_2SeO_3 (AlfaAesar 012585), respectively. Background ionic strength was 7.735 mmol/L KCl, the calculated equivalent to nutrient solution ionic strength (Chapter 3.3.3): 6500 µmol/L $Ca(NO_3)_2$ + 375 µmol/L K_2SO_4 + 325 µmol/L $MgSO_4$ + 400 µmol/L K_2HPO_4 + 8 µmol/L H_3BO_3 + 50 µmol/L $CaCl_2$ + 75 µmol/L $C_6H_5O_7Fe$ = 7.735 mmol/L ions (contributions smaller than 8 µmol/L were excluded). KCl was chosen instead of $CaCl_2$ because comparable experiments would be performed with additional ions (Chapter 4.4.4), such as sulfate and any Ca present would lead to precipitation (Chapter 3.3.1).

To avoid disproportional distortion due to different pipetting amounts of Se, the 18 Se solutions were prepared in double-deionized water in 25-mL glass flasks at double their appointed concentrations. Likewise, the KCl solution was prepared at 15.47 mmol/L (576.8 mg KCl, Merck, p.a. 1.04936.1000, in 500 mL double-deionized water). To achieve appointed Se and KCl concentrations, 15 mL of the respective Se solution and 15 mL of KCl were pipetted into a 50-mL glass beaker for each Se sorption solution, diluting each other into appointed concentrations. Solutions were homogenized, temperature and pH values were measured (WTW SenTix 81) and the solution was then evenly distributed, pipetting 10 mL each into the goethite and kaolinite patho-vessels (Figure 4.14b). The remaining 10 mL were transferred into a sampling patho-vessel and frozen at -20 °C until measurement. Kaolinite and goethite samples were shaken at room temperature for 24 h at 180 rpm; as the shape of patho-vessels allowed 10 mL of liquid to be homogenously shaken in an upright position.

After sorption, patho-vessels were centrifuged (Rotofix 32 A) at 40,000 rpm for 5 min. Because the patho-vessels are not guaranteed to withstand centrifuging, each patho-vessel was placed into a 50-mL centrifuge tube; however, none of the patho vessels broke. 9.5 mL of kaolinite sorption solution (9.25 mL of goethite sorption solution) were then sampled into labelled patho vessels with a pipette and frozen at -20 °C for later measurement. Sample analysis of Se was conducted with ICP-MS (Chapter 3.4.4).

4.4.3 Exp. B2 – Se desorption from kaolinite and goethite

The desorption of Se by means of the extraction step [Kulp & Pratt, 2004, Bacon & Davidson, 2007] as a desorption agent for easily exchangeable Se was intended to represent the amount of plant-available adsorbed Se.

With only 9.5 mL (9.25 mL) retrieved from the 10 mL initially added, directly adding desorption solution would dilute the remaining sorption solution and Se measured could not be monocausally identified; in the case of 5000 µg/L Se, by pipetting off 9.5 mL out of 10 mL, 250 µg/L Se would be measurable simply by filling up to 10 mL again from 0.5 mL. Therefore, a wash step was included, but not sampled for measurement; 9.5 mL of double-deionized water was pipetted onto the substrate, shaken for 1 min, centrifuged (40,000 rpm, 5 min) and pipetted off, thereby reducing even the highest-possible Se concentration in the event of no sorption from 5000 µg/L down to 12.5 µg/L Se.

For the actual desorption experiment, 9.5 mL of desorption solution was pipetted onto each of the previous sorption samples (desorption solution preparation: 500 mL of 0.1 mol/L K_2HPO_4 = 8.709 g, resulting pH = 9.23, brought down to pH 8.04 by 300 µL 9 mol/L HCl, Merck, p.a. 1.00317.2500) as described in soil extraction procedures [Bacon & Davidson, 2007]). Again, samples were shaken for 24 h at 180 rpm at room temperature. Samples were then centrifuged (40,000 rpm, 5 min, Rotofix 32 A), 7 mL sampled and frozen at -20 °C for later measurement. Sample analysis of Se was conducted with ICP-MS (Chapter 3.4.4) with a dilution factor of 10 due to high K_2HPO_4 sample salt load.

4.4.4 Exp. B3 – ion competition of Se sorption onto kaolinite and goethite

With the following experiment, competitive adsorption of selenite and selenate was determined for each of the major oxy-anion constituents of the nutrient solution used in experiment A3 (Chapter 3.3.3). The nutrient solution contained ca. 700 µmol/L of sulfate, ca. 12,000 µmol/L of nitrate and ca. 400 µmol/L of phosphate. For comparison between these ions within the sorption experiments and semi-comparison of sorption experiments to the nutrient solution, 750 µmol/L of each of the three main nutrient ions, nitrate, sulfate and phosphate, was separately added to the previous sorption set-up for competition sorption experiments as KNO_3 (Merck p.a. 1.05063.0500), K_2SO_4 (Merck, p.a. 1.05153.0500) or KH_2PO_4 (Merck, p.a. 1.04873.1000), respectively. Experiments were carried out with the same ionic strength of 7.735 mmol KCl for kaolinite and goethite in the presence of 750 µmol of either competing ion sulfate, nitrate or phosphate. Appointed Se concentrations (0, 10, 25, 50, 100, 250, 500, 1000, 2500 and 5000 µg/L) were again pipetted from 1000 mg/L stock solutions of $Na_2SeO_4 \cdot 10H_2O$ (VWR BDH Prolabo 302113L) and Na_2SeO_3 (AlfaAesar 012585).

Similar to the sorption experiment described above, for every batch of 20 samples, 0.5 g of either milled kaolinite (10 samples) or goethite (10 samples) were weighed (Sartorius 1712MP8) into labelled 20-mL patho vessels (Böttger 08-313-1001). Again, Se solutions were prepared in 25 mL glass flasks at double their appointed concentrations in double-deionized water. The KCl-oxy-anion solution was also prepared at double the appointed concentration in a 500-mL glass flask (15.47 mmol/L KCl = 576.8 mg KCl; 1500 µmol/L oxy-anion: 75.83 mg KNO_3, 130.70 mg K_2SO_4, 130.64 mg KH_2PO_4). To achieve appointed concentrations of Se, oxy-anions and KCl, 15 mL of the respective Se solution and 15 mL of the KCl-oxy-anion solution were pipetted into a 50-mL glass beaker for each Se sorption solution, diluting each other into appointed concentrations.

Again, these were homogenized, temperature and pH values were measured (WTW SenTix 81) and solutions were then distributed into a koalinite sample (10 mL) and a goethite sample (10 mL). The remaining 10 mL (the initial reference sample) were transferred into a sampling patho-vessel and frozen at -20 °C until measurement. Kaolinite and goethite samples were shaken at room temperature for 24 h at 180 rpm, centrifuged (40,000 rpm, 5 min, Rotofix 32 A) and sampled (5 mL). Samples were frozen at -20 °C until analysis of Se was conducted with ICP-MS and analysis of nitrate, sulfate and phosphate was conducted by IC (Chapter 3.4.4).

4.5 Results on Se adsorption onto kaolinite and goethite

Although preliminary experiments are usually not discussed in great detail, because they are generally not conducted under the exact conditions as the subsequent experiments (i.e. ionic strength being slightly different), they are included in the results in this case, as they have implications for the later discussion concerning the wash step and upscaling of experiments. Furthermore, sorption results in the presence of nutrient solution of Exp. C (Chapter 6.5c) are also included.

4.5.1 Preliminary tests

As shown in Table 4.2, pH values were measured with an accuracy of ±0.3 and values for blank, kaolinite and goethite converged consistently to 5.5, 6.8 and 7.8, respectively. The calculated proton concentration difference, which establishes between both minerals and double-deionized water (in equilibrium with CO_2 at 1 atm) at $pH_{init.}$ 5.5 was found to be nearly proportional to surface area, with a factor of 13.5 between surface areas of kaolinite and goethite and a factor of 12.9 - 13.2 between their respective changes in $c(H^+)/m^2$.

As shown in Figure 4.15a, when plotted against two logarithmic axes, selenate sorption showed a nearly linear trend, with goethite about twice as active as kaolinite. This means that both parameters – Se concentration in the solution and the amount of substrate added and, therefore, available sorption surface area – were nearly equally exponential in their influence. The highest adsorption in this study occured at the lowest available surface area (50 mg $\widehat{=}$ 0.47 m^2 kaolinite, 6.32 m^2 goethite) and the highest added Se concentration (5000 µg/L), leading to a load of 154 and 504 mg/kg Se on goethite and kaolinite, respectively.

Table 4.2: pH values for kaolinite (K) and goethite (G) in double-deionized water (Bl) after shaking for 24 h at 180 rpm and pH-calculated proton concentration change normalized per substrate surface area (pH was measured 5 times, 3 times with SentixV, 2 times with Sentix41)

ID	Kaolinite [g]	Goethite [g]	Surface [m^2]	pH-Mean [-]	SD [-]	c(H$^+$) [mol/L]	Δc(H$^+$)/surface [mol/m^2]
Bl 1	x	x	x	5.5	0.3	$3.162 \cdot 10^{-6}$	x
Bl 2	x	x	x	5.5	0.2	$3.162 \cdot 10^{-6}$	x
K 1	0.4942	x	x	6.8	0.2	$1.585 \cdot 10^{-7}$	$6.466 \cdot 10^{-9}$
K 2	0.5040	x	4.7376	6.8	0.4	$1.585 \cdot 10^{-7}$	$6.466 \cdot 10^{-9}$
G 1	x	0.4972	62.8461	7.7	0.3	$1.995 \cdot 10^{-8}$	$5.000 \cdot 10^{-10}$
G 2	x	0.5081	64.2239	7.8	0.3	$1.585 \cdot 10^{-8}$	$4.899 \cdot 10^{-10}$

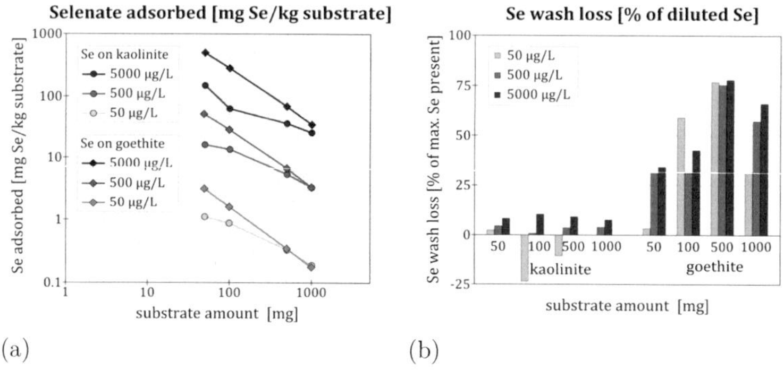

Figure 4.15: Se adsorption and wash results for preliminary sorptions tests
a: Selenate adsorption dependency on Se concentration and substrate amount compared for kaolinite and goethite
b: Measured Se loss through the wash/dilution step before desorption; with 0.5 mL and 0.75 mL in kaolinite and goethite samples remaining after sorption solution sampling are diluted by factor (20 and 13.3) to 10 mL with double-deionized water, Se content above 100 % is considered Se desorption

Results from the wash solution (Figure 4.15b) are normalized to 100 % of the Se expected to be found due to dilution of the remaining sorption solution (only 9.5 and 9.25 mL of the original 10 mL were sampled to avoid substrate loss, leaving 0.5 and 0.75 mL of residual Se in the sorption solution to be diluted to 10 mL). Figure 4.15b shows that no Se recovery of more than 100 % was found and this is interpreted to indicate no Se desorption via the wash step, which means that for further experiments, the wash solution step was not analyzed. However, although wash loss was below the maximum Se concentration calculated for dilution, it was shown to be 4 - 20 times higher for goethite than for kaolinite. This is attributed to pipetting error, as goethite was found to bind more water than kaolinite after centrifuging and the extraction of 9.25 mL was not always precisely possible without pipetting substrate as well. Moreover, analytically, only values for 5000 µg/L Se are above blind due to high analytically required dilution, Therefore, the factor of 3 - 7 compared to kaolinite can be explained by experimental and analytical conditions.

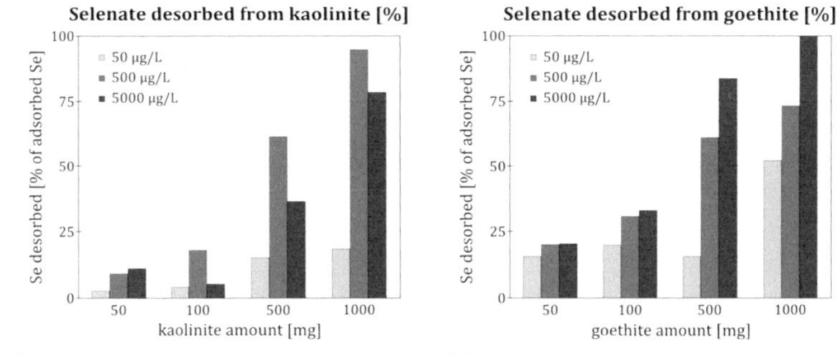

Figure 4.16: Se desorption results for preliminary sorptions tests as percentage of the previously adsorbed Se
a: desorption from kaolinite
b: desorption from goethite

None of the desorption percentages of previously adsorbed selenate exceed 100 % (Figure 4.16). There is a clear trend, however, with desorption lowest (below 25 %) for the lowest substrate amounts (50 - 100 mg kaolinite and 50 mg goethite) and increasing for greater substrate amount until reaching >75 % for 1000 mg kaolinite and goethite for desorption of 500 - 5000 µg/L Se.

Analytical difficulties in the reducing step of the HG-FIAS sample preparation procedure due to high concentrations of phosphate (from the 0.1 mol/L K_2HPO_4) required a change of analytical method and further experiments were analyzed by ICP-MS (Chapter 3.4.4).

4.5.2 Exp. B1 – Se sorption onto kaolinite and goethite

Se sorption onto kaolinite and goethite in this experiment differed between the substrates (Figure 4.17). While kaolinite showed logarithmic curvature in both adsorption isotherms for selenite and selenate, suggesting that the surface could reach its maximum load at approximately a 3 - 5-fold concentration of Se, Se sorption onto goethite was completely linear with Se concentration increase, suggesting little usage of its surface capacity.

Adsorption also differed between the anions selenite and selenate. Loss of Se from the solution due to sorption onto kaolinite decreased from 62 to 26 % for increasing selenite concentrations from 0 to 10,000 µg/L Se compared to decreasing from 74 to 40 % for increasing selenate concentrations from 0 to 10,000 µg/L Se (Figure 4.18). So while selenate appeared to have a higher affinity to kaolinite (30 - 47 % higher adsorption load) than selenite did, the opposite was true for goethite.

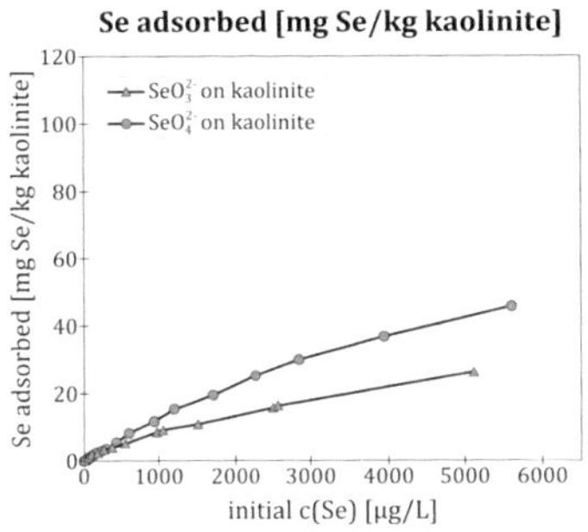

(a)

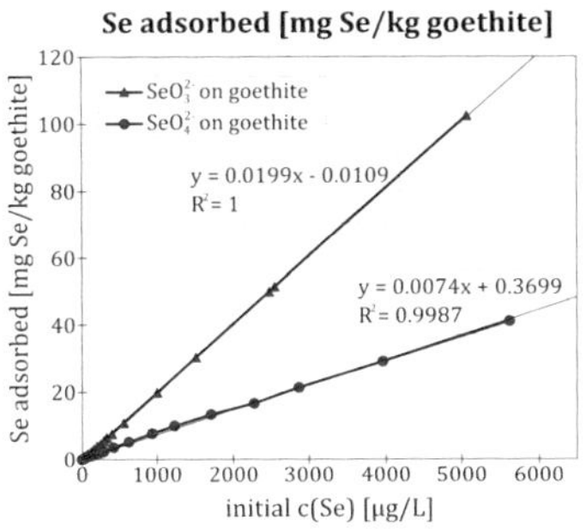

(b)

Figure 4.17: Se adsorption results of 0 - 5000 µg/L Se as $Na_2SeO_4^{2-}$ or $Na_2SeO_3^{2-}$
a: onto kaolinite
b: onto goethite

As shown in Figure 4.18b, selenite in solution was nearly completely (96 - 97 %) adsorbed onto goethite ($\geq$ 97 % when $\geq$ 500 µg/L Se was added), while only 43 - 36 % of the selenate in solution was adsorbed onto the goethite surface. There is little point in defining a coefficient for the difference in selenate and selenite sorption onto goethite, as the total adsorption of selenite to goethite indicates that selenite sorption onto the goethite surface was limited by the Se concentration provided in the solution.

Upon measuring the pH values before the sorption experiment, the Se-free initial solution blank had a pH value of 5.8 and 5.2 (Figure 4.19a), which corresponds well with the pH 5.6 predicted by PHREEQ-C (PHREEQ-C for Windows Version 2.18.00, wateq database) when the Se-free KCl-solution was equilibrated with typical air-content of 400 ppm of carbon dioxide [CO_2Now, 2014]. When adding Se to the solutions, pH values did not change significatly for the addition of selenate. With selenite, however, pH values increased with increasing Se addition, as shown in Figuere 4.19 a, and has been modelled by PHREEQ-C.

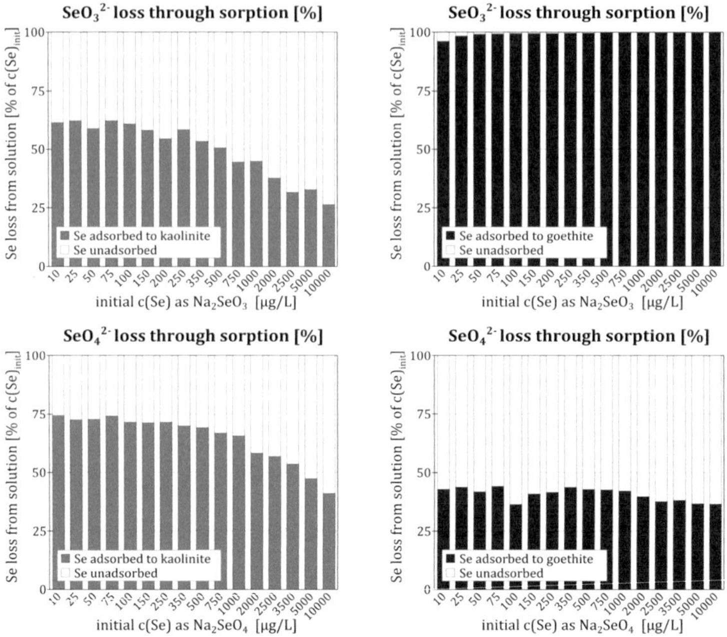

(a) (b)

Figure 4.18: Se loss from solution through sorption
a: after sorption onto kaolinite
b: after sorption onto goethite

Adsorption onto both mineral surfaces had a levelling effect on the pH regardless of the speciation. After sorption onto kaolinite, pH values rose to 6.6 - 6.8 when selenite was present, and 7.1 - 7.2 when selenate was the species being adsorbed (Figure 4.19b). After sorption onto goethite, pH values remained constant, regardless of the Se speciation between 7.9 and 8.1 (Figure 4.19b).

Concentrations of the prevalence of thermodynamically most abundant Se speciations in the modelled initial sorption solution are given in Figure 4.20. When adding Se as selenate, the most prevalent species by far in the modelled initial solution was the entirely deprotonated SeO_4^{2-}. Selenite, on the other hand, was present in all stages of protonization.

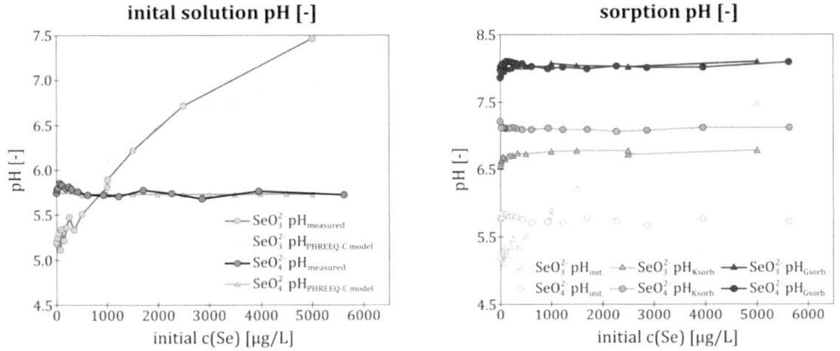

(a) (b)

Figure 4.19: pH values of Se sorption onto goethite and kaolinite
a: pH values of initital solutions as measured and calculated by PHREEQ-C
b: pH values of sorption solutions before ($pH_{init.}$) and after sorption of Na_2SeO_4 or Na_2SeO_3 onto kaolinite ($pH_{Ksorb.}$) and goethite ($pH_{Gsorb.}$)

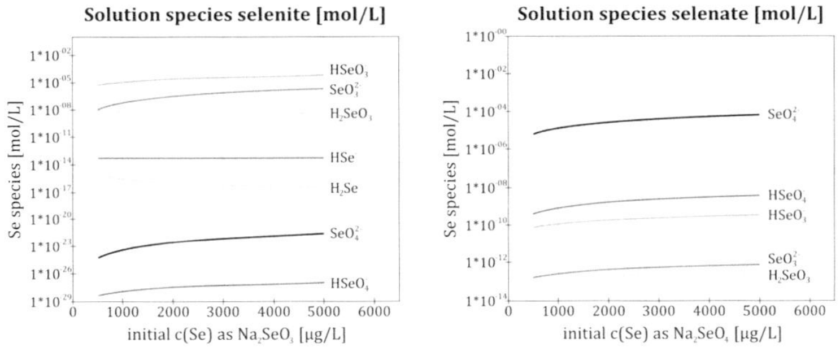

(a) (b)

Figure 4.20: PHREEQ-C modelling results for Se species in initital sorption solution.
a: after adding Se as selenite
b: after adding Se as selenate

4.5.3 Exp. B2 – Se desorption from kaolinite and goethite

Desorption profiles of Se from the surface to which it had previously been adsorbed to, was similar for both minerals (Figure 4.21). While desorption of selenate from kaolinite (85 - 96 %) was slightly more efficient than from geothite (71 - 94 %), desorption of selenite from kaolinite (76 - 92 %) was a little lower than that of selenate, while selenite desorption from goethite (77 - 87 %) was similar to that of selenate.

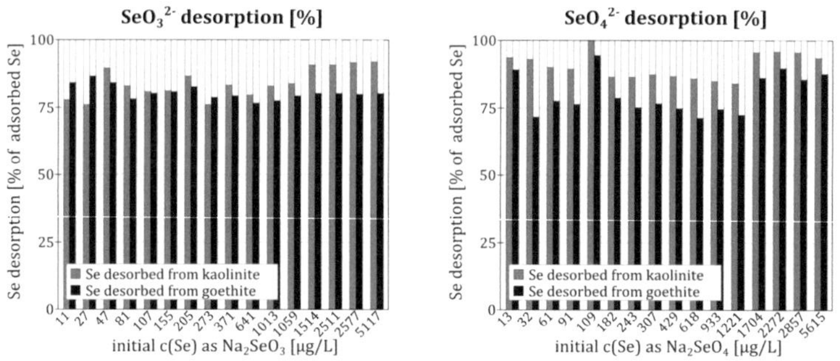

(a) (b)

Figure 4.21: Se desorption profiles from kaolinite and goethite using KH_2PO_4 as percentages of previously adsorbed Se.
a: desorption of selenite
b: desorption of selenate

4.5.4 Exp. B3 – ion competition and Se sorption

Compared to the adsorption of pure selenite onto kaolinite, the presence of 750 µmol/L N as nitrate or S as sulfate resulted in no significant change in selenite adsorption (Figure 4.22a). Selenite adsorption in the presence of nitrate was slightly improved compared to pure adsorption (98 - 138 %; $\bar{x}$ = 117 ±14) and in the presence of sulfate, barely influenced (87 - 124 %; $\bar{x}$ = 99 ±14). However, phosphate, as expected, had a great impact on selenite adsorption onto kaolinite.

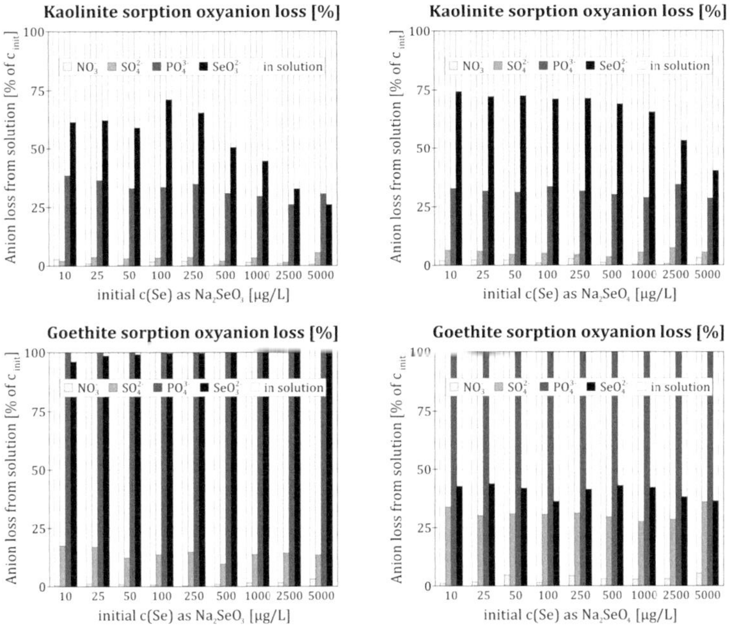

(a) (b)

Figure 4.22: Se adsorption onto kaolinite (top) and goethite (bottom) in the presence of 750 µmol of competing ions nitrate, phosphate, sulfate in comparison with pure Se adsorption and Se adsorption in the presence of nutrient solution (Exp. C, Chapter 6.5c)
a: adsorption as selenite
b: adsorption as selenate

In the presence of 750 µmol/L of P as phosphate, selenite adsorption was reduced to 12 - 38 % ($\bar{x}$ = 20 ±9). With the nutrient solution containing 400 µmol/L P rather than 750 µmol/L (= 53 % c(P)), the inhibition of selenite adsorption was explained in part (70 %) as phosphate-induced (Figure 4.22a).

Selenate adsorption, on the other hand, was inhibited by each of the ions to varying proportions (Figure 4.22b). While selenate adsorption was moderately inhibited (66 - 80 %; $\bar{x} = 72 \pm 5$) by 750 µmol/L N as nitrate, it was very strongly inhibited by the equivalent amount of S as sulfate (2 - 10 %; $\bar{x} = 5 \pm 3$) and P as phosphate (2 - 8 %; $\bar{x} = 6 \pm 2$). In the presence of nutrient solution (Chapter 6.5c), selenate adsorption was 22 - 42 % ($\bar{x} = 31 \pm 7$) compared to pure selenate adsorption and, therefore, inhibited less in combination of nitrate, phosphate and sulfate than with each ion individually (Figure 4.22b).

When comparing sorption on goethite between pure selenite solution and selenite solution in combination with other competing oxy-anions, Figure 4.22a clearly shows that all available selenite was always completely adsorbed, regardless of accompanying ions. This was true for all singly competing ions, as well as the nutrient solution. In all cases, goethite did not reach a maximum load.

Selenate adsorption onto goethite, on the other hand (Figure 4.22b), was inhibited by the presence of phosphate (4 - 12 %; $\bar{x} = 9 \pm 3$) and sulfate (4 - 15 %; $\bar{x} = 9 \pm 5$), with values similar to selenate adsorption onto kaolinite. While a maximum load was once again not reached, the equilibrium amount of selenate adsorbed onto the goethite surface was lower for sulfate and phosphate, but – as with adsorption onto kaolinite – selenate adsorption was not negatively influenced by nitrate, but rather increased (105 - 157 %; $\bar{x} = 118 \pm 24$). Selenate adsorption in the experiment with nutrient solution showed the highest adsorption (278 - 387 %; $\bar{x} = 283 \pm 75$).

Adsorption of the oxy-anions is shown as loss of nitrate, phosphate and sulfate from solution due to adsorption onto kaolinite and goethite in Figure 4.23. Consistent with its small influence on Se adsorption, nitrate loss from solution due to adsorption onto kaolinite was nearly insignificant (0.4 - 3.2 %; $\bar{x}$ = 1.3 ±0.8 and $\bar{x}$ = 1.85 ±1.0 for selenite and selenate, respectively). Loss of sulfate from solution was only slightly higher, with about twice as much (2 - 7.5 %; $\bar{x}$ = 3.2 ±1.2 and $\bar{x}$ = 5.4 ±1.1 for selenite and selenate, respectively). Phosphate loss, however, was quite significant (26.0 - 38.3 %; $\bar{x}$ = 32.6 ±3.8 and $\bar{x}$ = 31.5 ±2.0 for selenite and selenate, respectively), which was also consistent with the adsorption of selenite and selenate being affected by phosphate presence.

With goethite, again, loss of nitrate was nearly insignificant (0.2 - 5.5 %; $\bar{x}$ = 1.2 ±0.9 and $\bar{x}$ = 3.2 ±1.5 for selenite and selenate, respectively), albeit up to twice as high as with kaolinite. Loss of phosphate from all experimental solutions was always 100 % (Figure 4.23). Contrary to kaolinite results, loss of competing sulfate from solutution due to adsorption onto goethite was dependent on the prevalent Se species. Loss of sulfate in the presence of selenite was 9.5 - 17.5 %; $\bar{x}$ = 14.1 ±2.4, while in the presence of selenate it was 27.4 - 36.0 %; $\bar{x}$ = 30.9 ±2.6. This is consistent with its previously reported influence on selenite and selenite.

4.6 Discussion on Se sorption properties and ion competition

Apart from the obvious differences between the surface area of both minerals used, the main points of discussion for the adsorption of selenite and selenate on kaolinite and goethite address the difference in Se oxy-anion affinity for the substrates. Beyond that, the differences in competition posed by the anions sulfate, phosphate and nitrate are detailed in particular. The discussion concludes with the major points

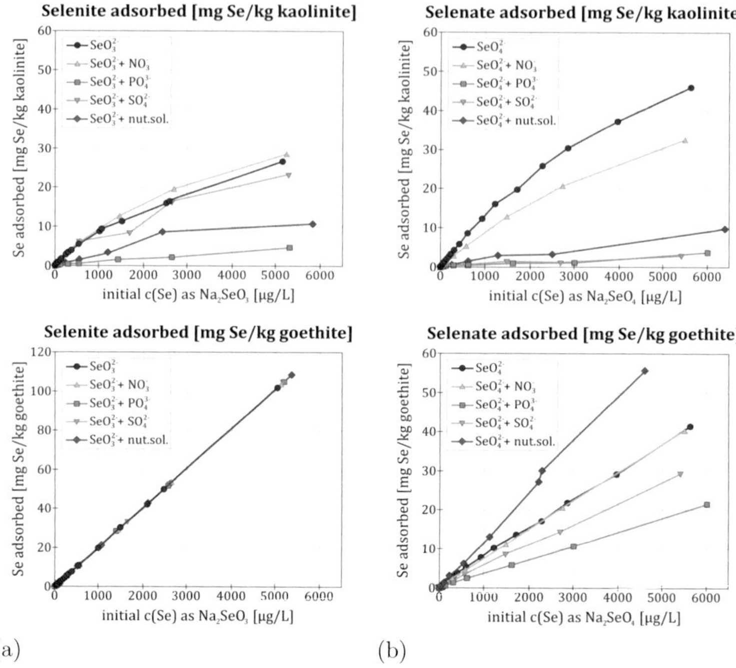

Figure 4.23: Loss of anion concentration from solution through adsorption onto kaolinite (top) and goethite (bottom)
a: adsorption of selenite
b: adsorption of selenate

required for the following biogeochemical model and the mass balance calculations.

As the greater goal of this study was primarily concerned with characterizing and quantifying processes in the Se cycle within the Critical Zone, the following experiments were not designed to fulfill requirements on which to base a detailed, numerical sorption complexation model. Moreover, such data on microscopic mechanisms of Se oxy-anion complexation on mineral surfaces has already been published in various other studies [Manceau & Charlet, 1994, Wijnja & Schulthess, 2000, Peak,

2006, Chan et al., 2009]. Therefore, only macroscopic observations are used to answer questions related to this study. Since this study's purpose was to combine sorption studies with plant uptake studies, ionic strength, pH value and the concentrations of competing ions used are specifically tailored to this goal and have not been used in previous studies.

4.6.1 Selenium complexation with kaolinite and goethite surfaces

While there is no dissent in the literature on selenite complexation, there is uncertainty about the details of selenate adsorption. It is widely accepted that selenate's type of complexation varies, depending on surrounding parameters, such as solution pH and ionic strength as well as substrate surface [Fernández-Martínez & Charlet, 2009]. Selenate is not the only oxy-anion which can bind both inner- and outer-spherically; arsenate and sulfate, for example, share this property as well [Catalano et al., 2008, Chan et al., 2009].

Selenate's behavior during the washing step (Figure 4.15b) shows no macroscopic evidence for selenate being bound as loosely as would be expected of an outer-sphere complex. Outer-sphere complexes are expected to be more susceptible to ionic strength variations than inner-sphere complexes because, according to the TLM (Chapter 4.5), background electrolyte ions are placed in the same plane as the outer-sphere complexes [Wu et al., 2000]. A solution of double-deionized water, therefore, would be expected to lead to a partial exchange of Se anions with OH^-. This was not observed (Figure 4.15 b) – neither for the adsorption onto kaolinite, nor for the adsorption onto goethite – for which inner-spheric adsorption of selenate has been shown [Hayes et al., 1987, Manceau & Charlet, 1994, Wijnja & Schulthess, 2000].

Moreover, desorption using phosphate would be expected to exchange nearly 100 % of the adsorbed Se, which is not the case (Figure 4.21). Rather, selenate desorption from goethite is similar to selenite desorption and selenate desorption from kaolinite is only slightly higher compared to selenite desorption. For this study, selenate is, therefore, considered to be adsorbed primarily as an inner-sphere complex, like selenite. This corresponds well with literature identifying selenate an intermediately bonding anion compared to the strongly bonding selenite, which has been found to bond prefferably as inner-sphere complex onto goethite [Wu et al., 2000, Wijnja & Schulthess, 2000].

4.6.2 Se-oxy-anion affinities to kaolinite and goethite

The main difference between selenite and selenate was that selenite has a very high affinity to the goethite surface, compared to selenate affinity to goethite as well as either Se species' affinity to kaolinite. While selenate has a slightly higher affinity to the kaolinite surface than selenite, its affinity to goethite is much lower (Figure 4.18) considering the vast difference in surface area.

With surface area (9.4 and 126.4 m^2 for kaolinite and goethite, respectively) so different, it comes as no surprise that the goethite surface was not saturated with Se or any of the other oxy-anions as opposed to kaolinite. The method used to describe surface activity in these experiments is a primitive one compared to the possibilities of obtaining values for pH_{PZC}, pH_{PZNPC} or pH_{IEP} [Sposito, 1998], [Appel et al., 2003] and gives no approximation for either value. However, this pH characterization of surface activity is representative for initial experimental conditions and sufficient for the purposes of this study.

As pH_{PZC} is an indicator of the pH range for a substrate's physisorption properties, pH_{PZC} values for goethite between 7.3 and 9.4 [Parks, 1990, Kosmulski, 2002, Appel et al., 2003] make both coulombic and chemical adsorption possible for the ions used in this experiment (pH 5.2 - 5.7). However, the final pH after adsorption, which is changed independent of Se concentration is explained, not through Se adsorption, but by the goethite surface acting as a base for the solutions protons. Thereby deprived of protons, the solution's pH value rises to 7.9 - 8.1. Due to the goethite sample's large surface area, this effect is overwhelmingly larger than the oxy-anion adsorption.

Values for kaolinite's pH_{PZC}, however, range from $pH_{PZC} = 3 - 5$, corresponding to the surface groups [Schroth & Sposito, 1997, Hur & Schlautmann, 2003] to $pH_{PZC} = 6 - 7.2$, corresponding to the amphoteric edge groups [Hur & Schlautmann, 2003, Tombácz & Szekeres, 2006]. The initial sorption solution pH of 5.2 - 5.7 means that coulombic attraction is plausible only for the surface groups, but not for the edge groups, which means that physisorption for Se oxy-anions in this environment is reduced and chemical adsorption must overcome the charge repulsion. This makes adsorption onto kaolinite less effective than onto goethite. With final pH values of kaolinite sorption at 6.6 - 6.8 and 7.1 - 7.2 for selenite and selenate addition, respectively, there appears to be strong adsorption activity from the edge groups rather than the surface groups, while selenite may also adsorb onto surface groups, raising the final pH by 0.5 less compared to selenate adsorption.

The results of this study, although not focused on complexation modelling, are considered indicative of inner-sphere complexation of both selenite and selenate ions onto the kaolinite surface. These findings are interesting because selenate, in general, shows a lower affinity to surfaces than selenite [Wu et al., 2000, Chan et al., 2009, Fernández-

Martínez & Charlet, 2009]. For adsorption onto kaolinite, however, the higher amount of selenate adsorption compared to selenite adsorption can be attributed to steric properties, since the selenite molecule is larger than the selenate molecule and, therefore, occupies more space on the surface (Figure 4.8). Unfortunately, while quite a few studies have been published on Se adsorption onto goethite [Manceau & Charlet, 1994, Rietra et al., 2001, Wijnja & Schulthess, 2002, Rovira et al., 2008, Das et al., 2013], there is little data on specific Se adsorption onto kaolinite [Fernández-Martínez & Charlet, 2009].

The finding that selenite has such an overwhelming affinity for the goethite surface is not new [Hayes et al., 1987, Lo & Chen, 1997, Chan et al., 2009, Fernández-Martínez & Charlet, 2009]. In general, there is little information on direct comparison between selenite and selenate with regard to their varying affinities to goethite because in modelling, it is incorporated as a constant derived from chemical equilibria [Wu et al., 2000, Rovira et al., 2008]. Sorption rates have been shown to be higher for selenite than for selenate [Rovira et al., 2008]. The fact that selenite and selenate exhibit different kinetic processes and that selenite binds as a bidentate, rather than a mono-dentate complex like selenate, might provide an answer to the high selenite-goethite affinity [Zhang & Sparks, 1990, Lo & Chen, 1997, Su & Suarez, 2000, Fernández-Martínez & Charlet, 2009].

For quantification purposes of this study, it can be concluded that when adsorption remains in a linear range, which means the substrate surface is not maximally loaded or sorptive transport is not hampered by extreme dilution, selenite's great affinity for kaolinite and goethite leads to 61 % and 99 % selenite loss from solution, respectively, while selenate loss from solution is 72 % for kaolinite and 42 % for goethite.

4.6.3 Desorption

Desorption with phosphate is considered an effective way to desorb selenite [Rajan & Watkinson, 1976, Kulp & Pratt, 2004]. As shown in Figure 4.21, desorption ranges of selenite and selenate are comparable for both kaolinite and goethite. Similar to the values of desorption in the literature [Singh et al., 1981], with the exception of outliers, overall desorption is between 71 and 96 % with little indication for a difference between kaolinite desorption for either Se oxy-anion to be found. However, selenate appears to desorb slightly better from kaolinite than from goethite with the use of phosphate. This has previously been decribed [Singh et al., 1981] and can be explained by the mono-dentate inner-sphere complex of selenate, which is more easily exchanged than the bidentate inner-sphere complex of selenite.

KH_2PO_4 was used to determine the easily exchangeable fraction of Se, which in a natural environment might, therefore, be accessible to plants. Findings of these experiments show that around 75 % of both Se species can be desorbed by ion exchange with phosphate – even when it is adsorbed to goethite. Therefore, soils rich in ferric oxides or hydroxides might contain high concentrations of Se even if the soil solution contains only traces. This Se might then unintentionally be mobilized by large quantities of fertilizer containing phosphate. It remains to be seen, however, whether the exchangeable Se determined by this study truly also equals plant-available Se in environments without excess phosphate.

4.6.4 Oxy-anion competition

Natural waters contain many different ions. In an optimally fertilized environment of plant cultivation, adsorption of selenite and selenate are affected by oxy-anions, such as nitrate, phosphate and sulfate. Predictions for oxy-anion affinities for adsorption onto surfaces are difficult, since they are highly dependent on charge distribution between central atoms and ligand atoms of both the substrate and the anion. This and the affinity for protonization of both reaction partners are dependent on pH value and ionic strength [Hiemstra & van Riemsdijk, 1995].

As this experiment was able to demonstrate, nitrate only affects selenate adsorption onto kaolinite and is otherwise not a factor in Se adsorption, neither in selenate adsorption onto goethite, nor in selenite adsorption onto either kaolinite or goethite. This is not surprising, as nitrate's structure was very different from that of both Se oxy-anions, since it had a much larger radius and also much lower electronegativity difference (Figure 4.8), making it much less likely to adsorb onto surfaces than Se oxy-anions. Moreover, nitrate is only onefold negatively charged, while sulfate is two- and phosphate is threefold negatively charged. In published literature, nitrate was listed with lower general adsorption affinity [Neal, 1995] and early studies already proved nitrate to be adsorbed much less than competing sulfate or phosphate [Kinjo & Pratt, 1971]. As expected, sulfate poses strong competition for selenate, and only little competition for selenite, while phosphate is strongly inhibitative of Se oxy-anion adsorption in general. This is not surprising as it was shown to be a good desorption agent for both Se oxy-anions as well [Balistrieri & Chao, 1990, Geelhoed et al., 1997, Duc et al., 2003]. This is probably due to its high charge-to-radius ratio coupled with the adsorption structure of a bidentate complexation to the surface [Hiemstra & van Riemsdijk, 1995, Geelhoed et al., 1997].

Direct comparison between both mineral substrates is difficult due to their large difference in surface area. Previous literature has described clay minerals as having low retention properties towards anionic species such as selenite and selenate [Duc et al., 2003]. When comparing selenate adsorption onto kaolinite and goethite, however, this is not necessarily true, as adsorption per area was higher on kaolinite than on goethite, despite the factor 13 in surface area difference in favor of goethite. In contrast to that, considering selenite's high affinity for goethite and goethite's large surface area, it comes as no surprise that selenite adsorption is not inhibited by anything present in the experimental nutrient solution, although phosphate shows at least the same affinity for the goethite surface as selenite and is adsorbed to 100 %. Alhough this behavior is not directly transferrable to conditions of natural soils, published literature suggests that selenite adsorption onto ferric oxides is a great source of Se retention, rendering selenite bio-unavailable in the presence of ferric oxides and hydroxides [Bar Yosef & Meek, 1987, Howard, 1977, Fernández-Martínez & Charlet, 2009].

As a final point of discussion, the effects of combined oxy-anion competition in the nutrient solution are of great interest, since this is, arguably, the most realistic kind of natural adsorption environment. The effects of multiple ion competition cannot simply be described by the sum of single-ion effects. In the case of selenite adsorption onto kaolinite, 70 % of the nutrient solution's effect could be ascribed to the phosphate share. However, with nitrate and sulfate barely influencing selenite adsorption as single-ions, this poses the question where the other 30 % of selenite inhibition comes from. The reasons for this were not explored in detail and can, therefore, only be surmised to be related either to other constituents of the nutrient solution, i.e. borate or ferric citrate, or to other parameters such as pH value and experimental set-up, which was slightly different for the other experiments. The fact that adsorption of selenite onto goethite is by far most efficient in the nutrient solution environment might be ascribed to bridging of more than one ion onto certain Fe-sites, although there is only sparse infomation on this process [Hingston et al. 1971]. It is also possible that Se forms complexes in the nutrient solution and that goethite, a known scavenger [Bar-Yosef & Meek, 1987], shows a greater affinity toward sorption of complexes than to selenate alone.

4.6.5 Conclusions for biogeochemical mass balance modelling

The results of this study have direct implications for the biogeochemical model with which characterization and mass balance of the Se compartments and transfer processes within the Critical Zone can be described. Although mathematical correlations do not necessarily follow chemical or physical properties or processes, the purpose of this model is to define an adequate description of correlations and quantification of processes. Therefore, the model proposed for an experimental system must be oriented on the basis of the data produced by this system and all assumptions and boundary conditions are based on macroscopic observation.

The sorption processes describe one-half of the pre-defined model transport pathways for Se, which was compartmentalized into "soil" and "soil solution". Conclusions from this study for this aspect of the Critical Zone are:

1. selenite and selenate both adsorb as inner-sphere complexes with species- and substrate-dependent affinities;
2. the small kaolinite surface size allows isotherm modelling;
3. adsorption onto the large goethite surface size can be simplified as linear adsorption;
4. as plants require time to germinate, the sorption equilibrium can be modelled as pre-installed before plant-induced effects occur.

5 Biogeochemistry – Se uptake competition of minerals and plants

In this chapter, the experimental set-ups of Chapters 3 and 4 are combined to study synergetic effects of the uptake of Se into plants in the presence of competing mineral surfaces. As introductions to both plant uptake of Se (Chapter 3.1) and adsorption of Se onto minerals (Chapter 4.1) have already been presented in their respective chapters, this chapter will provide a brief overview of consequences of previous experimental results for the combined experiments.

5.1 Theoretical implications of previous results

Although Exp. A3 was combined with Exp. B3 in a closed plant-box system (Figure 5.1), Se adsorption onto kaolinite or goethite preceded the uptake of Se from the nutrient solution by the rice plants in this set-up due to germination time. As selenite and selenate have been reported to equilibrate with i.e. goethite in 25 min [Su & Suarez, 2000], Se adsorption onto the substrate already reached an equilibrium while the plant had no contact with the nutrient solution yet.

Since the same nutrient solution was used as in Exp. A3, no toxicity effects were expected for concentrations up to 2500 µg/L Se (Figure 5.1) and uptake of Se was believed to be similar to Exp. A3, with similar partitioning between shoots and roots for selenate and selenite, respectively. Because, however, kaolinite exhibited Se adsorption between 23 and 75 % of solution-Se and goethite adsorbed between 25 and 50 % of solution selenate, experimental set-ups were modified to provide higher Se-concentrations (0 - 10,000 µg/L Se) than those in Exp. A3 (0 - 2500 µg/L). Nonetheless, selenite affinity to goethite was observed to be so great in Exp. B3 that no Se uptake by the rice plants was expected.

5.2 Experimental set-up

As shown in Figure 5.1, in the fourth plant-box experiment of this study, plants were grown in the presence of a nutrient solution identical to the one described in Chapter 3.3.3. Additionally, mineral substrate of kaolinite and goethite was added in the exact weight-volume ratio used in Exp. B1 and B3 (Chapters 4.4.2 and 4.4.4). For a plant-free nutrient solution sorption reference, which was directly comparable to Exp. B1 and B3, each plant-box had a sorption bottle-equivalent, with the identical solution compared to the plant-box and the same amount of mineral substrate (Figure 5.1).

5.2.1 Exp. C – plant-box rice cultures with Se-nutrient solution & mineral sorption

A 1000 mg/L Se stock solution was prepared in 10 mL of double-deionized water ($Na_2SeO_4 \cdot 10H_2O$, VWR BDH Prolabo 302113L and Na_2SeO_3, AlfaAesar 012585). Ninety rice caryopses were sterilized according to the procedure described in Chapter 3.3.1. 8.5 g of either kaolinite or goethite were weighed (Sartorius 1712MP8) into 10 plant-box sets (Sigma Aldrich, 2 x V8505 & 1 x C0667) and 10 250-mL Schott glass bottles.

The substrate-containing plant-box sets were prepared (Figure 5.1a) with a specially designed Eppendorf tube tray holder for 9 x 1.5-mL tubes (1.5 mL Eppendorf Safe-Lock Microcentrifuge Tube, 0030120086). Plant-box sets, substrate-containing 250-mL Schott glass bottles (all pre-labelled autoclave-proof), 90 1.5-mL Eppendorf tubes, 250 mL 0.7 % phytoagar, 10 500-mL Schott glass bottles with 400 mL nutrient solution each (preparation given in Chapter 3.3.3), a 100-mL measuring cylinder and 250 mL of double deionized water were autoclaved.

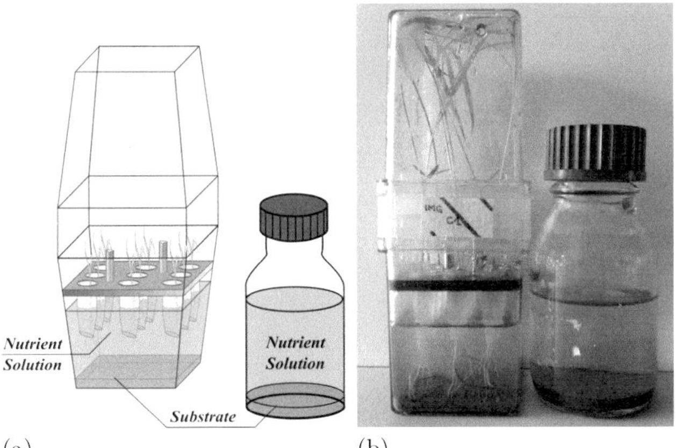

(a) (b)

Figure 5.1: Se-nutrient solution experiments contained 9 rice plants growing for 19 days, sprouting first in Se-free agar in Eppendorf tubes before reaching 170 mL of Se-spiked nutrient solution (0 - 10,000 µg/L Se) in the presence of competing Se adsorption by kaolinite or goethite mineral surface.
a: schematic experimental set-up
b: photo taken on 03.06.2014

After autoclaving, 1.5-mL reaction tubes (Eppendorf, Hamburg) were placed into a tray-holder under the cleanbench and filled to the brim with the still-liquid 0.7 % phytoagar. Only after the agar had solidified, were the lid and 0.5 cm from the bottom of each reaction tube cut off using sterilized scissors. Each tube was then loaded with one sterilized caryopse.

Due to autoclaving, nutrient solution compositions varied slightly from bottle to bottle (variation within one experimental batch 0.8 - 13.3 % , $\bar{x} = 4.9$ %). To ensure that plant-box experiments were directly comparable to their plant-free sorption references, the initial nutrient solutions were prepared in the autoclaved 500-mL Schott glass bottles as follows (Figure 5.2a), before being aliquoted into the plant-box, sorption reference bottle and sample vessels.

Under the cleanbench, Se standard was added to the nutrient solution in the autoclaved 500-mL Schott glass bottles to achieve specified concentrations (0, 20, 50, 100, 200, 500, 1000, 2000, 5000, 10 000 µg/L Se). Solutions in the 500-mL bottles were shaken to homogenize. Autoclaved plant-boxes were disassembled under the cleanbench and Eppendorf trays removed. From the nutrient solution in the 500-mL Schott glass bottles, 170 mL of nutrient solution was filled into each lower half of the boxes and an equal 170 mL filled into its respective autoclaved reference glass bottle. Two times 15 mL of the remaining solution in the 500-mL Schott glass bottles were sampled for subsequent analysis into 20-mL patho-vessels (Böttger 08-313-1001) – 15 mL were frozen for later IC-measurement, the other 15 mL were acidified with 50 µL of HNO_3 (VWR 20429.320 p.a. sub-boiled) and stored at 4 °C.

For plant-box assembly, each Eppendorf tray was loaded with 9 rice-bearing Eppendorf tubes and placed into the plant-box with the lower half of the tubes submerged in nutrient solution before closing the plant-box. The 250-mL Schott reference glass bottles were screwed shut and not shaken. Plant-boxes and reference bottles were kept under the same climate chamber conditions as described in Chapter 3.3.3, keeping the boxes and bottles closed for 19 days until harvest. In total, four experimental set-ups were conducted: two with kaolinite as the sorption substrate and Se addded as selenite and another two with goethite as the sorption substrate with Se added as selenate. Each experimental set-up was conducted only once without further replicates.

(a) (b)

Figure 5.2: Photos of Exp. C preparation and harvest.
a: preparation of goethite experiments under sterile conditions of the cleanbench; goethite in disassembled plant-boxes and 250-mL reference bottles (red lids) and prepared nutrient solution in 500-mL bottles (blue lids), caryopses planted into Eppendorf tubes with bottom cut off on the right, patho vessels for sampling in the front (photo taken on 08.05.2014)
b: harvest of kaolinite experiment (photo taken on 13.04.2014)

5.2.2 Sample preparation and analytical methods

Plants were harvested (Figure 5.2b) as described in Chapter 3.4.1, shoot and root bulk samples were digested using the protocol described in Chapter 3.4.2 and subsequent analysis of plant-Se with HG-FIAS was conducted according to Chapter 3.4.3. 2 x 15 mL of nutrient solution was sampled from the plant boxes as well as from the reference sorption bottles into 20-mL patho-vessels (Böttger 08-313-1001) – 15 mL were frozen for later IC-measurement, the other 15 mL were acidified with 50 µL of HNO_3 (VWR 20429.320 p.a. sub-boiled).

As shown in Figure 5.3a, the substrate of plant-boxes and reference bottles was filtered through cellulose filters (Schleicher & Schuell, 595, No. 311610), washed once with double-deionized water (comparable to the wash-step before desorption experiments in Chapter 4.4.3) and dried on the filters in an oven at 40 °C for 48 h. As shown in Figure 5.3b, samples were then filled into 20-mL patho-vessels (Böttger 08-313-1001) for storage, later analysis and desorption experiments.

(a) (b)

Figure 5.3: Photos of substrate sample preparation after harvest.
a: filtration of substrate from nutrient solution onto cellulose filtres (photo taken on 14.04.2014)
b: sampling of dried material from the filtre into a patho-vessel (photo taken on 16.04.2014)

5.2.3 Desorption experiments

From the filtered sorption substrate material, desorption samples were prepared similarly to those described in Chapter 4.4.3. 0.5 mg of substrate sample were weighed into 20-mL patho-vessels (Böttger 08-313-1001) and 10 mL of desorption solution (0.1 mol/L K_2HPO_4, prepared as in Chapter 4.4.3) was pipetted onto it. Samples were shaken for 24 h at 180 rpm at room temperature. Samples were then centrifuged (40,000 rpm, 5 min, Rotofix 32 A), 7 mL sampled into 20-mL patho-vessels (Böttger 08-313-1001) and frozen at -20 °C for later measurement. Sample analysis of Se was conducted with ICP-MS (Chapter 3.4.4) with a dilution factor of 10 due to a high K_2HPO_4 sample salt load.

5.2.4 Scanning Electron Microscopy analysis (SEM)

From each of the four experimental runs of Exp. C, to which 5000 µg/L Se had been added, sorption substrate material of plant-box experiments and their respective sorption reference bottles were taken to the LEM (KIT) for SEM analysis. Samples were coated in platinum by vapor deposition to enable static discharge and measured with a LEO 1530 Gemini at 10 keV operating voltage and beam focal distance of 5.1 - 6.3 mm.

5.2.5 Synchrotron-based analysis

Synchrotron radiation is generated when accelerated electrons are redirected by the use of bending magnets. Specialized beamlines using this tangentially escaping radiation are installed around an electron storage ring, in which accelerated electrons are circulated [Newville, 2004]. Methods used for this study included X-ray absorption near edge spectroscopy (XANES) of speciation-dependent Se absorption (SUL-X beamline, ANKA, Karlsruhe) and fluorescence mappings (FLUO beamline, ANKA, Karlsruhe) of plant tissue Se content.

XANES scans detail how X-rays are absorbed by an atom at energies near (±30 eV) the core-level binding energies of an atom and is strongly sensitive to formal oxidation state and coordination chemistry of the absorbing atom [Newville, 2004]. X-rays (ca. 500 eV to 500 keV) as part of the synchrotron radiation spectrum, are absorbed when the incoming radiation energy equals the binding energy of an electron [Newville, 2004]. The electron is thereby removed from its quantum level [Newville, 2004]. If the incoming radiation energy exceeds that of the electron binding energy, the excess energy is transferred to a photo-electron which is then ejected from the atom. Using Lambert-Beer's law, X-ray absorption can be described and calculated [Newville, 2004].

$$\mu(E) \cdot t = ln(I_0/I_{trans}) \tag{5.1}$$

with

$\mu(E)$	as the absorption coefficient
t	sample thickness
I_0	as the intensity of incident radiation
I_{trans}	as the intensity transmitted through the sample

A so-called "absorption edge" is observed when the incident X-ray has an energy equal to the binding energy of a core-level electron. Core-level binding energies for electrons are specific to each atom, which makes XANES an element-specific measuring method [Newville, 2004]. After absorption, the atom is in an excited state with a core electron level vacant and an emitted photo-electron. This excited state decays within femtoseconds after absorption by either fluorescence or the Auger Effect [Newville, 2004]. Fluorescence describes the process of a higher-level electron filling the vacant electron-hole by emitting the defined excess energy as light [Newville, 2004]. In the case of Se, the Kα energy denoting the photon energy released during an electron transition from the L to the K shell is at 11.2 keV [Tokunaga et al., 1994]. The Auger Effect, on the other hand, describes the process in which the emitted excess energy of a higher-level electron filling the vacant electron-hole is not emitted as light, but rather transferred to another electron, which is then ejected from the atom, leaving it with two electron vacancies [Newville, 2004]. These two processes do not interfere with X-ray absorption, but are directly proportional to it [Newville, 2004]. Therefore, the Auger Effect or fluorescence effect can be used to determine the absorption edge instead of directly measuring transmission, when, for example, sample thickness allows only little transmission or element concentration is very low and there is little difference between incident and transmitted radiation.

In the case of this study, fluorescence was the underlying process used both for XANES measurements and fluorescence mappings. The difference between the former and the latter was only one of analytical set-up and interpretation. While XANES use fluorescence in its proportionality to absorption when induced with a radiation spectrum of ±30 eV around the absorption edge (for Se: 12.658 keV), fluorescence mappings are carried out with a fixed incident radiation energy and fluorescence of different atoms is distinguished by the energy increments of the emitted fluorescence (for Se: 11.222 keV).

Samples with sufficiently high concentrations, such as reference materials can be measured in transmission mode (measuring absorption of the sample by loss of intensity between ionization chamber 1 and 2, Figure 5.4). In this case, absorption is calculated as $\ln(\text{abs}) = \ln(I_0/I_{\text{trans}})$. Samples with low concentrations can be more accurately measured in fluorescence mode (measuring fluorescence of the sample, which is proportional to the absorption; Figure 5.4). In this case, absorption is calculated as $I_{fluo}= I_{fluo}/I_0$. Additionally, each measurement (reference or sample alike) can be energy-calibrated by measuring absorption of an elemental Se standard (measuring standard absorption loss of intensity between ionization chamber 2 and 3, Figure 5.4), with which all scans can be calibrated and aligned to the energy of the element's absorption edge E_0.

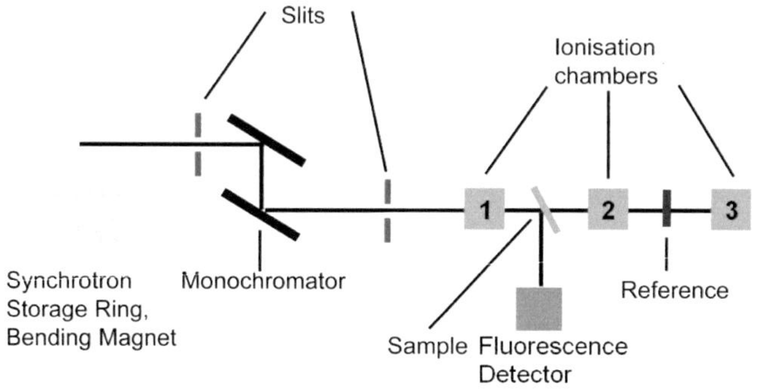

Figure 5.4: Scheme of the experimental setup for XANES analysis at a synchrotron facility ([Newville, 2004, Diener, 2012])

Sample preparation

From each of the kaolinite sorption plant box experiments, one rice plant was selected per box for measurements at the synchrotron facility ANKA, Karlsruhe. Harvested along with the other plants, these were not freeze-dried. After rinsing under once-deionized water and 2 h of drying in lab conditions, each plant was placed in a separate 15-mL centrifuge tube (VWR 525-0149) with the lid only loosely screwed on. To avoid any changes of Se speciation in the plant tissue, samples were not freeze-dried, but kept in a desiccator for 4 weeks to air-dry prior to analysis to avoid Se-loss and preserve plant structures during the drying process.

XANES analysis

XANES were measured at the SUL-X beamline (ANKA, Karlsruhe). This beamline was equipped with a 7-element fluorescence detector (e2v; Resolution: <140 eV at 5.9 keV, 1000 cps and <310 eV at 5.9 keV, 100.000 cps), three ionization chambers for absorption measurement (Oxford Instruments, IC-Plus type; active length 5 cm, Kapton windows 6 µm thick), a CCD detector (Photonic science; 80 x 120 mm^2, Fiber optic 3.46:1, 2048 x 2048, 16 bit dynamic, readout time 3.3 s – 21 s) and an optical microscope (TSO Spezialoptik, resolution 2 µm).

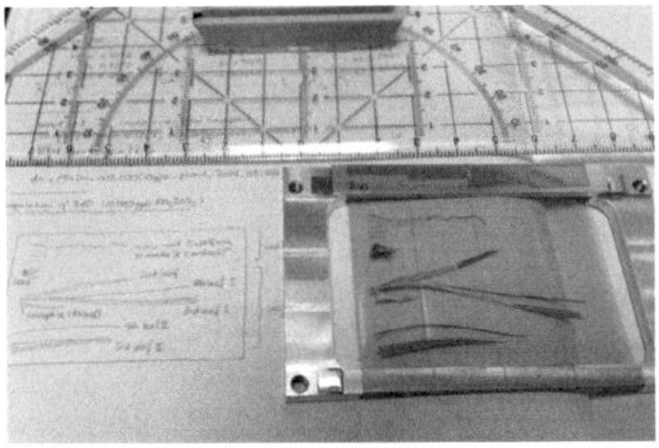
(a)

(b)

Figure 5.5: Sample preparation for synchrotron-based analysis of air-dried plant samples
a: XANES plant sample on kapton-tape (right) and schematic sample documentation (left) (photo taken on 22.05.2014)
b: Ralph Steininger closing the vacuum chamber for test XANES measurements of a plant sample using a cryo-cell (photo taken on 20.05.2014)

Table 5.1: XANES measurement parameters: stepsize and acquisition time of pre-edge, absorption edge and post-edge

scan	parameter	pre-edge I	pre-edge II	absorption edge: 12.658 keV	post-edge
quickscan	range	-75 to -20 eV		-20 to +20 eV	+20 to +200 eV
	acq. time	1 s		1 s	1 s
	stepsize	1 eV		0.5 eV	1.5 eV
absorption spectrum	range	-100 to -50 eV	-50 to -25 eV	-25 to +25 eV	+25 to +144 eV
	acq. time	1 s	1 s	1 s	1 s
	stepsize	5 eV	2 eV	0.5 eV	1.5 eV

Seleno-methionine (Sigma Aldrich S3132-100 MG), Na_2SeO_3 (AlfaAesar 012585) and $Na_2SeO_4 \cdot 10H_2O$ (VWR BDH Prolabo 302113 L) were prepared as reference materials by finely grinding the material in an agate mortar and spreading it over the sticky side of kapton tape (M. Newville, 2014, personal communication; IDE-13 beamline, APS, Argonne). Eight layers of this prepared tape were glued on top of each other. All reference samples (for later linear combination fitting) were measured under vacuum in transmission against the Se(0) standard prepared as a press pellet with 8 mg of elemental Se (Merck 1.07714.0050) in 100 mg of cellulose powder (Sigma Aldrich C8002-1KG).

Plant samples were fixed on kapton tape and placed on a metal frame (Figure 5.5a). Using an optical microscope, regions of interest (labelled 01 - 09) on the sample were calibrated relative to reference points on the sample frame and microscope images were taken of each spot. The sample holder was then installed in the measuring chamber and put under vacuum.

To locate hotspots of Se on the sample for XANES measurements, Se profiles across plant sections (measured across the point from -0.5 mm to +0.5 mm in 50 steps for 1 s) in the regions of interest (ROI) were measured at 12.661 keV. Then, absorption spectra (details below) of 2 - 8 points (labelled a - h) of the profile with sufficient Se content were measured for each ROI. Each of these points was measured twice and scans were merged for greater accuracy. XANES measurements were carried out under vacuum and with short acquisition times (ca. 7 min) to minimize beam-induced redox reactions. Although references were measured in transmission mode, in many cases, Se tissue concentration was too low to measure in transmission mode, so absorption was measured in fluorescence mode for all plant samples.

To determine measurement conditions best suited to avoid beam-induced damage, one sample was measured with three set-ups. The same sample was measured first using atmospheric conditions, second, in a chamber under vacuum and third, using a cryo-cell cooled with liquid nitrogen in the vacuum chamber (Figure 5.5b). Beamtime measuring current during measurement varied by 5 - 15 mA, the beam spot on the sample was 30 x 50 µm in diameter and the fluorescence detector was between 60 and 80 mm away from the sample. Table 5.1 details how absorption spectra and quickscans, in which the motor changing the light energy does not pause while fluorescence is detected, were measured.

Data was processed as ".spec" data files using the Athena program of the Demeter software package (Demeter 0.9.13, B. Ravel, M. Newville). In ".spec-files" of the SUL-X beamline at ANKA, data for these calculations are given in the columns described in Table 5.2. For normalization, the pre-edge was set to -90 to -45 keV before E_0 and the post-edge was set to +90 to +245 keV after E_0 in the calibration scans. Samples were normalized to a pre-edge of -90 to -45 keV before E_0 and the post-edge was set to +90 to +190 keV after E_0 in the sample scans. Using linear combination fitting, proportional weighting of reference scans in all combinations was calculated to match the merged sample scans across -30 eV to +30 eV around the absorption edge.

Table 5.2: Available data columns for the processing of SUL-X beamline data

data	column No.
energy	1
C_{fluo} (fluorescence counts)	56
I_0 (intensity in ionization chamber 1, incoming intensity)	59
I_{abs} (intensity in ionization chamber 2, intensity after transmission through sample)	60
I_{ref} (intensity in ionization chamber 3, intensity after transmission through reference)	61
calculated absorption	62
calculated reference absorption	63

High resolution µXRF mapping

In an unsuccessful first attempt at µXRF (micro X-ray fluorescence) mapping of Se concentration in plant material, samples were prepared using the Immuno-fluorescence fixation method into paraffin [Lützelschwab, 1990] by using a fixation agent to fix all biochemical compartments to the cell structure and then replacing – in incremental steps – tissue water by ethanol, ethanol by xylene and xylene by paraffin. Using a cutter, thin sections of the paraffin-embedded sample were manually sliced to ca. 0.5 - 1 mm thickness. However, the fluorescence measured could not be assigned to plant tissue structures, leading to the conclusion that the Immuno-fluorescence fixation method did not preserve cell structures and Se was distributed uniformly across plant tissue and in some cases bled into the paraffin embedding material.

Therefore, fluorescence of the complete air-dried plant samples used previously for XANES measurements was mapped at an excitation energy of 22 keV at the FLUO beamline at ANKA, Karlsruhe. This beamline was equipped with a double multilayer monochromator (W-Si multilayers in 2.7 nm period), CRL: 2×10^9 ph/s at 17 keV (5 µm x 2 µm), 1 ionization chamber, Si(Li)-energy dispersive detector (Oxford Instruments), HPGe- High Purity Germanium detector (Princeton Gamma-Tech (PGT)) and a SiMCD-Vortex Silicon Multicathode Detector. Two undiluted press pellets of bulk plant material (Figure 5.6a) with known Se concentrations (measured with ICP-MS after acid digestion) from Punjab, India were prepared to calibrate Se concentration in the mappings: *Brassica juncea* root with 186 mg/kg Se DW and *Brassica juncea* leaf with 931 mg/kg Se DW. Plant samples were glued to slide frames (Butch, 28565348) between kapton tape to ensure as plane a surface as possible (Figure 5.6b). This was necessary, because during mesh analysis mode (meshs = 2D grids of measuring points), sample distance from detector could not be varied between measuring points.

(a) (b)

Figure 5.6: Sample preparation for synchrotron-based μ-XRF analysis of air-dried plant samples
a: bulk press pellets for quantitative concentration calibration prepared on kapton tape fastened to a magnetic sample holder under the light of a binocular (photo taken on 04.06.2014)
b: fluorescence mapping samples between kapton tape on a slide frame for a maximum of planar surface (photo taken on 06.06.2014)

Using a binocular, regions of interest were calibrated relative to an easily identifiable marker point. The sample slide frame was then installed in the measuring chamber. To locate exact positions of Se on the sample, Se fluorescence profiles (at 11.2222 keV for $k_{\alpha 1}$) across plant sections (measured across the point from -1 mm to +1 mm in 50 steps for 1 s) in the ROIs were measured at an excitation energy of 22 keV. Then, meshs were measured across ROIs. Beam spot on the sample was 16 x 11 µm (polycapillary focus) and acquisition time was 3 - 4 sec. Data was analyzed using the PyMCA (version 3.9.5) software and mappings were created with Surferplot 6 (Golden Software).

5.3 Results

As these experiments are a synthesis of the previous experiments Exp. A2 - A4 (Chapters 3.3.2, 3.3.3, 3.3.4) and B1 - B3 (Chapters 4.4.2, 4.4.3, 4.4.4), results for parameters of plant health, such as plant mass, shoot height and heights of the second leaf as well as Se content results of rice shoots and roots from Exp. C, are directly compared with the respective results from Exp. B3 to determine additional effects of the possible presence of sorption in Exp. C. Furthermore, results for the adsorption of Se from the nutrient solution onto kaolinite or goethite during Exp. C are shown in Figures 4.22 in Chapter 4.5.4.

5.3.1 Plant growth of *Oryza sativa*

Shoot growth height of plants in the kaolinite experiment (Figure 5.7a) showed no correlation to added Se between initial Se concentrations of 20 - 1000 µg/L Se, with heights of 102 - 87 % and 90 - 81 % of the untreated shoot height for Se added as selenite or selenate, respectively. With the addition of 2000, 5000 and 10,000 µg/L Se as selenite, however, shoot height decreased to 75, 41 and 28 % of untreated shoot height, respectively. With the addition of 5000 and 10,000 µg/L Se as selenate, shoot height decrease was not quite as pronounced at 71 and 59 % of untreated shoot height, respectively.

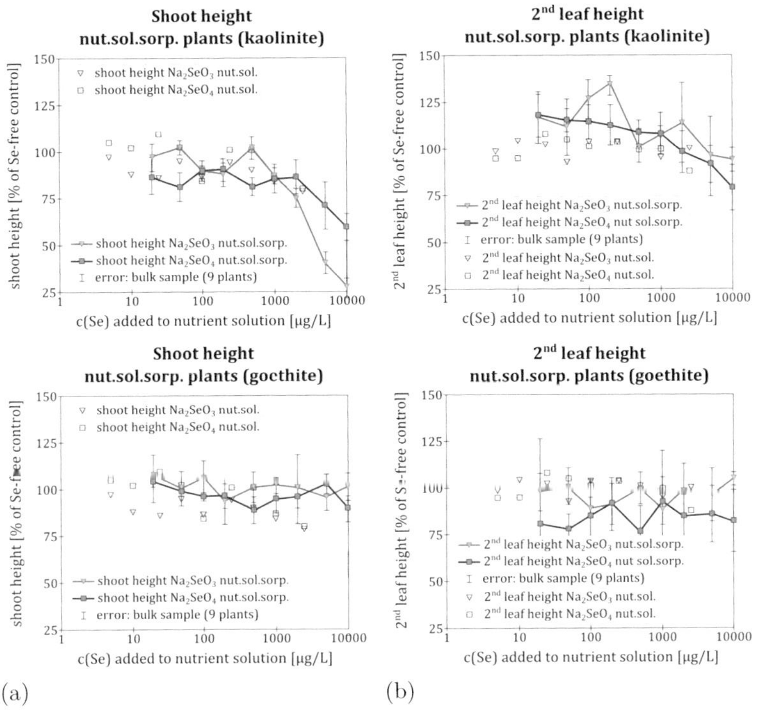

Figure 5.7: Plant growth expressed as percentage of Se-free blank plants for experiment C in competition with kaolinite (top) and goethite (bottom) adsorption
a: total shoot height on harvest day (Day 16)
b: total length of 2nd leaf on harvest day (Day 16)

A similar decrease in shoot height was also found for the height of the second leaf; however (Figure 5.7b), this was less pronounced. Heights for the second leaf in the kaolinite experiment showed an overall slightly decreasing trend from 116 - 94 % and 118 - 79 % compared to the untreated plants for selenite and selenate, respectively.

Shoot height of plants in the goethite experiment (Figure 5.7a) showed no correlation with added Se across the initial concentrations 20 - 10,000 µg/L Se, with heights of 107 - 93 % and 103 - 88 % compared to the untreated plant height for selenite and selenate, respectively. Height of the second leaf in the goethite experiment (Figure 5.7b) also showed no correlation with added Se, with heights of 105 - 90 % and 92 - 77 % compared to the untreated plant height for selenite and selenate, respectively. This corresponds well with the lack of correlation with added Se concentration for the experiments without substrate, since any fluctuations observed in shoot height or height of the second leaf were within 1s-error marging, as discussed in Chapter 3.5.3.

5.3.2 Selenium content in *Oryza sativa*

Quality measures for experiment C show that harvested plant yield was 85.6 % of transferred seedlings. Digestion standard retrieval was 83 % (±13) and drinking water standard retrieval during HG-FIAS measurement was 121 % (±9).

Selenium concentrations measured in the rice plants of Exp. C in the presence of kaolinite increased from 1.4 to 80 mg/kg DW for shoots (Figure 5.8a) and from 3.1 to 444 mg/kg DW for roots with the addition of 20 - 10,000 µg/L Se as selenite. With the addition of Se as selenate, plant-Se in the presence of kaolinite increased from 0.8 to 1080 mg/kg DW for shoots (Figure 5.8b) and from 0.1 to 402 mg/kg DW for roots.

Selenium concentrations measured in the rice plants in the presence of goethite increased from 0.0 to 1.9 mg/kg DW for shoots (Figure 5.8a) and from 0.0 to 9.9 mg/kg DW for roots with the addition of 20 - 10,000 µg/L Se as selenite. With the addition of Se as selenate, plant-Se in the presence of goethite increased from 0.7 to 347 mg/kg DW for shoots (Figure 5.8b) and from 0.2 to 84 mg/kg DW for roots.

As in Exp. A3, overall plant-Se content increased with additional Se in the nutrient solution in all set-ups of Exp. C. For experiments in the presence of kaolinite, addition of 100, 1000 and 10,000 µg/L Se as selenite led to 8.5, 65 and 80 mg/kg DW shoot-Se and 14, 152 and 444 mg/kg DW root-Se (Figure 5.8a). However, there was one exception. Shoot-Se content peaked at 117 mg/kg DW with the addition of 2000 µg/L Se as selenite. For Se solution concentrations higher than that, 5000 and 10,000 µg/L Se, shoot-Se decreased to 86 and 80 mg/kg DW, respectively.

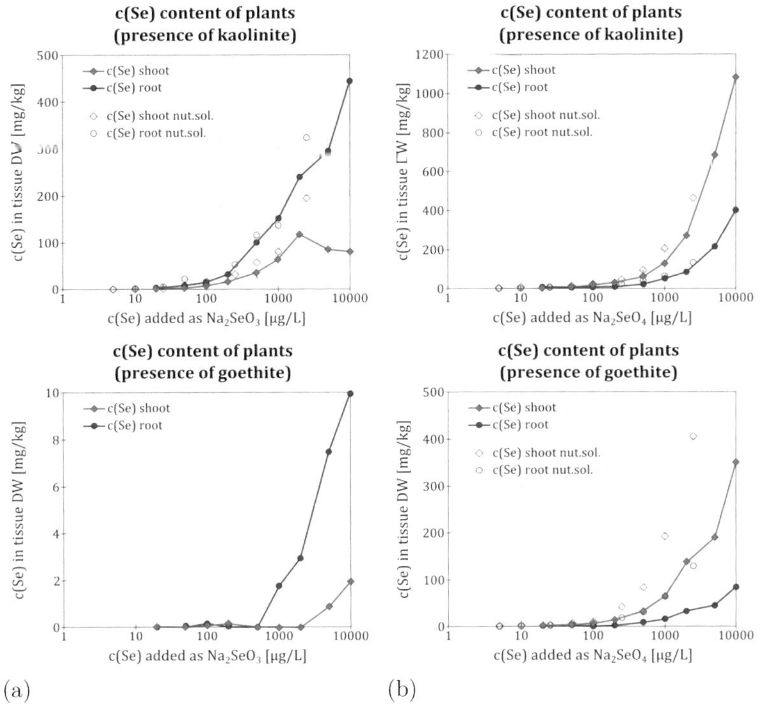

Figure 5.8: Se content in roots and shoots of Exp. C; Se-uptake in competition with adsorption by kaolinite (top) and goethite (bottom) compared with data from Exp. A3 without sorption (data from Exp. A3 is omitted for selenite uptake in the presence of goethite, because Se-uptake in Exp. C was too small for comparison)
a: addition of Se as Na_2SeO_3
b: addition of Se as Na_2SeO_4

For selenate experiments in the presence of kaolinite, additions of 100, 1000 and 10,000 µg/L Se led to 12, 127 and 1080 mg/kg DW shoot-Se and 3.2, 49 and 402 mg/kg DW root-Se (Figure 5.8b), respectively. Here, higher Se concentration led to higher Se content in the plant.

In the presence of goethite, very little Se was taken up into the plant when it was added as selenite compared to the uptake of Se as selenate. Plant-Se in the shoots was only measurable for the addition of 5000 and 10,000 µg/L, resulting in 0.9 and 1.9 mg/kg DW, respectively. In the roots, Se-content was measurable at additions of Se greater than 500 µg/L and increased with the concentration of added Se from 1.8 mg/kg DW with the addition of 1000 µg/L Se to 10 mg/kg DW with the addition of 10,000 µg/L Se.

Selenate additions in the presence of goethite showed an increase in plant-Se with added Se concentrations. Additions of 100, 1000 and 10,000 µg/L Se as selenate led to 5.2, 63 and 347 mg/kg DW of shoot-Se and 1.1, 15 and 84 mg/kg DW of root-Se, respectively.

As with Exp. A2, A3 and A4, partitioning of selenate predominantly to the shoots and selenite predominantly to the roots was observable in the results of Exp. C as well, regardless of the addition of substrate, as shown in Table 5.3. Accumulation factors for the experiments in the presence of kaolinite with the addition of Se as selenite were 22.2 ($\pm$3.1) in shoots in concentrations of 20 - 2000 µg/L Se; then decreased to 6.6 and 2.9 for initial Se concentrations of 5000 and 10,000 µg/L Se, respectively. Selenium accumulation factors in the roots when added as selenite decreased with increasing initial Se concentrations from 138 to 23.9 for initial Se concentrations of 20 to 10,000 µg/L Se, respectively. Accumulation factors of Se in plants when added as selenate in the presence of kaolinite was 20.5 ($\pm$4.7) and 9.1 ($\pm$2.9) for shoots and roots, respectively. Therefore, Se accumulation in roots when added as selenite was an average of 5.7 ($\pm$1.8) times higher than in shoots in the

Table 5.3: Fresh weight accumulation factors (AF) of selenate and selenite into shoots in Exp. C and Exp. A3 as calculated from measured Se content in the nutrient solution (AF [-] = c(Se) medium [mg/L] / c(Se) plant [mg/kg])

	Exp. $C_{kaolinite}$				**Exp. $C_{goethite}$**				**Exp. A3**			
added	AF-Na_2SeO_3		AF-Na_2SeO_4		AF-Na_2SeO_3		AF-Na_2SeO_4		AF-Na_2SeO_3		AF-Na_2SeO_4	
c(Se)	shoot	root	shoot	root	shoot	root	shoot	root	shoot	root	shoot	root
[µg/L]	[-]	[-]	[-]	[-]	[-]	[-]	[-]	[-]	[-]	[-]	[-]	[-]
5									21.5	80.3	27.2	21.7
10									26.7	81.6	16.2	8.0
20	24.2	138	10.6	3.8	0.3	2.9	19.6	8.3	26.5	99.2	20.9	9.7
50	22.7	148	22.6	11.4	0.2	0.7	11.3	13.1	31.6	212	26.7	12.7
100	22.7	118	27.1	10.3	26.1	153	25.0	9.8	27.0	105	19.0	10.2
200	26.8	132	24.4	8.5	18.2	0.6	34.5	6.5	29.0	94.1	41.1	19.9
500	18.2	156.6	17.4	13.5	0.1	4.8	34.3	15.7	33.6	117	40.0	21.7
1000	18.1	98.8	24.4	9.6	0.4	545	29.8	12.4	30.7	84.8	53.4	25.5
2000	22.9	55.3	18.7	9.9	1.4	554	35.7	14.0	22.1	57.8	40.9	22.5
5000	6.6	30.4	19.8	9.0	21.2	595	49.3	23.0				
10,000	2.0	23.0	10.7	5.5	26.1	460	40.5	14.6				

presence of kaolinite; accumulation of Se as selenate in shoots was an average of 2.4 (±0.7) times higher than in roots.

For the low concentrations available for plant uptake in the goethite experiments with selenite, uptake, when measurable, showed a high accumulation factor in the roots (153 - 596), but an accumulation comparable to the kaolinite experiments in the shoots (18.2 - 26.1). In the selenate experiments, Se accumulation, in the presence of goethite was 31.1 (±11.3) and 13.0 (±4.8) into shoots and roots, respectively. Accumulation of Se as selenate in shoots was, therefore, an average of 2.6 (±1.1) times higher than in roots in the presence of goethite.

When compared to the results of Exp. A3, where there was no sorption interference, accumulation of Se as selenite in roots was an average of 3.7 (±1.2) times higher than in shoots and accumulation of Se as selenate in shoots was an average of 1.9 (±0.3) times higher than in roots. Partitioning of the accumulation of Se in plant tissue was, therefore, more pronounced in Exp. C.

Table 5.4: Plant-Se of Exp. C given as percentage of plant-Se of Exp. A3 for respectively comparable Se additions ($c(Se)_{Exp.\ C}/c(Se)_{Exp.\ A3}$)

	Na_2SeO_3				**Na_2SeO_4**			
	kaolinite		goethite		kaolinite		goethite	
added c(Se)	shoot	root	shoot	root	shoot	root	shoot	root
[µg/L]	[%]	[%]	[%]	[%]	[%]	[%]	[%]	[%]
20	50.3	46.5	0.1	0.1	33.1	28.7	33.1	18.6
50	57.0	31.8	0.0	0.0	69.7	55.6	19.9	29.0
100	73.4	44.3	1.1	0.7	112.3	82.8	57.2	33.5
200	51.1	48.7	0.4	0.0	51.1	35.4	34.0	9.3
500	50.5	59.5	0.0	0.0	61.7	69.6	40.3	26.2
1000	53.0	70.9	0.0	1.0	55.2	65.3	33.1	24.0
2000	61.5	64.3	0.0	0.9	55.6	55.5	33.8	25.0
mean	56.68	52.27	0.22	0.38	62.65	56.14	35.90	23.66
mean-SD	8.43	13.38	0.39	0.46	24.58	18.97	11.22	7.81

Under the assumption that plant-Se standard deviation was comparable to that of Exp. A3 (standard deviation across all plant-Se of Exp. A3: 47.9 % for the addition of 10 - 50 µg/L Se, 30.3 % for the addition of 100 - 500 µg/L Se and 19.8 % for the addition of 1000 - 2500 µg/L Se, Figure 3.16, Chapter 3.5.3), Se-uptake into plants in the kaolinite experiments was comparable to that of Exp. A3 and within that margin of error. However, these high standard deviations mask that additional substrate had an effect on the mean values of plant-Se. With the exception of shoot-Se with the addition of 100 µg/L Se added as selenate to the kaolinite experiment, these were all below mean values of Exp. A3 (Table 5.4), in which there was no sorption substrate. For kaolinite experiments, plant-Se with the addition of either selenate or selenite was between half and two-thirds of the Exp. A3 plant-Se (52.27 - 62.65 %) and for goethite experiments, plant-Se was between a quarter and a third of the Exp. A3 plant-Se when added as selenate. Plant-Se when added as selenite for goethite experiments was below 1 % of comparable Exp. A3 plant-Se.

5.3.3 Root toxicity

In kaolinite-substrate experiments, addition of Se as selenite produced toxicity symptoms in the rice plants at the addition of 5000 and 10,000 µg/L Se. As shown in Figure 5.9, in the case of the addition of 5000 µg/L Se, all plant roots grew through the agar in the Eppendorf tube, but no further than 0.5 cm into the nutrient solution-containing the Se. In the case of the addition of 10,000 µg/L Se, also shown in Figure 5.9, only the main root (one root per plant) grew through the agar of the Eppendorf tube until reaching the nutrient solution. For all plants of this plant-box, this main root showed brown-spotted discoloration and discontinued growth at less than 3 mm into the nutrient solution. All other roots remained stunted at less than 1.5 cm in length and also exhibited brown discoloration despite not being directly in contact with the Se. Experiments with selenate did not show signs of toxicity, nor did the selenite experiments with goethite, since adsorption of goethite led to strong depletion of Se from the nutrient solution (Chapter 5.3.4).

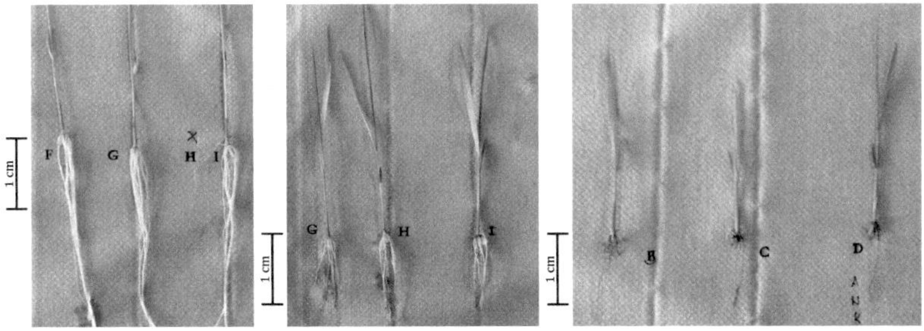

Figure 5.9: Comparison of root growth for rice plants grown in contact with Se as Na_2SeO_3 in nutrient solution, in Se concentrations of 2000 µg/L Se (left), 5000 µg/L Se (middle), 10,000 µg/L Se (right)

5.3.4 Adsorption influence of kaolinite and goethite

As expected in light of the previous results of Exp. B1 - B3, Se was not only taken up by plants, but also adsorbed onto the additional substrates kaolinite and goethite.

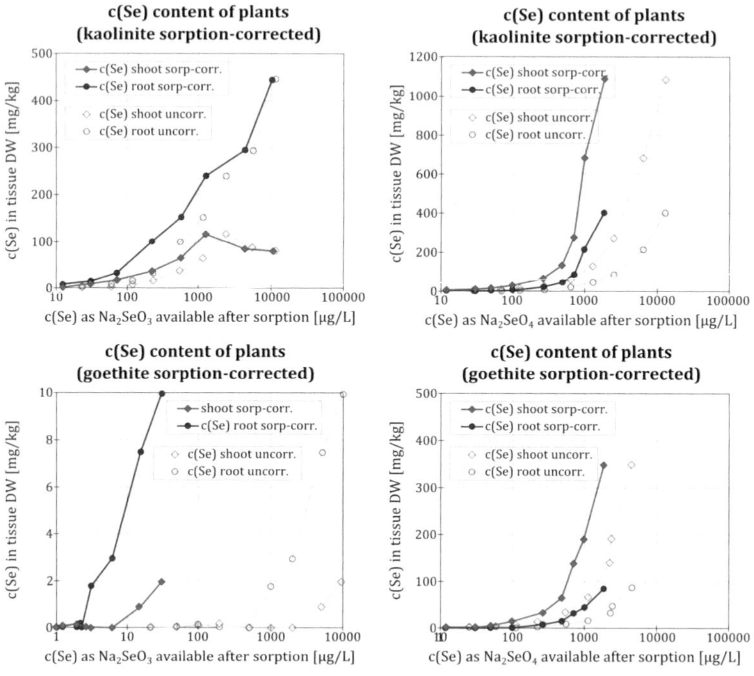

(a) (b)

Figure 5.10: Se content in plant tissue with sorption competition by kaolinite (top) and goethite (bottom); initial Se concentration corrected by adsorption determined from bottle refercence experiments compared to the uncorrected initial Se concentration
a: addition of Se as Na_2SeO_3
b: addition of Se as Na_2SeO_4

Under the assumption that the same amount of Se was adsorbed by the substrate in the plant boxes than in the reference bottles, initital Se content of the nutrient solution was corrected by the amount of Se adsorbed by the substrate to determine Se content at the time the plant roots reached the nutrient solution (Figure 5.10). Calculated factors ($c(Se)_{tissueDW}/c(Se)_{nut.sol.}$) of sorption-corrected plant-Se uptake compared to the uncorrected uptake reflected that substrate influence.

In the presence of kaolinite, sorption-corrected plant uptake of Se was calculated to be $\bar{x}$ = 119 % ±7 and $\bar{x}$ = 112 % ±4 of the uncorrected uptake for selenite and selenite, respectively. In the presence of goethite, sorption-corrected plant uptake of Se was calculated to be $\bar{x}$ = 27150 % ±9608 and $\bar{x}$ = 112 % ±4 of the uncorrected uptake for selenite and selenite, respectively. This made goethite the greater influence on either selenite or selenate uptake, with selenite uptake affected disproportionately. Kaolinite had a similar influence on uptake of either Se oxy-anion.

With the main difference in experimental set-up being the additional ions in the nutrient solution, adsorption results of Se in Exp. C differed from those in Exp. B1 (Figure 5.11a,b). Although kaolinite adsorption (Figure 5.11a) did not reach a maximum – as was the case in Exp. B1 – adsorption of selenite compared to selenate showed only small differences, unlike results of Exp. B1, where selenate adsorption to kaolinite was higher than selenite adsorption. Moreover, adsorption in the presence of nutrients in Exp. C. was much lower compared to Exp. B1, where the background ion concentration was comprised solely of KCl.

Compared to Exp. B1, in which adsorption was studied in the presence of KCl solution rather than a nutrient solution, adsorption onto kaolinite was reduced to $\bar{x}$ = 17 % ±5 and $\bar{x}$ = 39 % ±8 for selenate and selenite, respectively. Adsorption onto goethite (Figure 5.11b) was increased to

$\bar{x}$ = 126 % ±31 compared to Exp. B1 and remained unchanged at $\bar{x}$ = 99 % ±9 in the presence of nutrients for selenate and selenite, respectively.

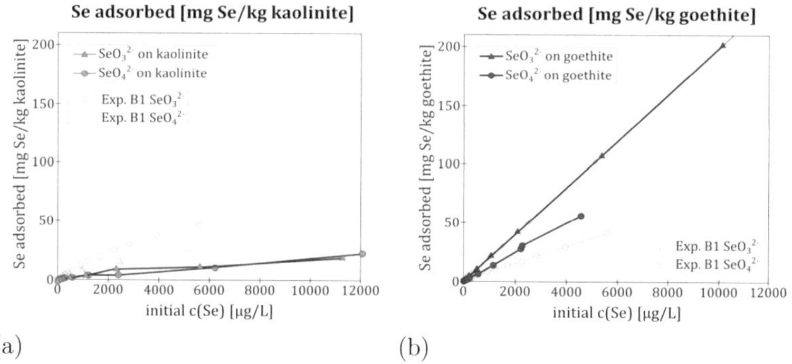

Figure 5.11: Se adsorption from nutrient solution as Na_2SeO_3 and Na_2SeO_4
a: onto kaolinite
b: onto goethite

As with the results in Exp. B1, this substrate influence was visible in the Se loss of the solution. Again, assuming experimental conditions between plant boxes and reference bottles were comparable, Se loss due to kaolinite adsorption in Exp. C (Figure 5.12 (top)) was $\bar{x}$ = 16 % ±5 and $\bar{x}$ = 10 % ±3, while plant uptake only accounted for $\bar{x}$ = 12 % ±4 and $\bar{x}$ = 6 % ±3 of the Se loss for selenite and selenate, respectively.

This is much lower than in Exp. B1, where Se loss due to kaolinite adsorption was between 26 and 62 % for selenite and 40 to 74 % for selenate. Selenium loss due to goethite adsorption in Exp. C (Figure 5.12 (bottom)), however, showed greater similarity to that of Exp. B1. Sorption accounted for $\bar{x}$ = 98 % ±2 and $\bar{x}$ = 57 % ±4, while plant uptake only accounted for $\bar{x}$ = 0.5 % ±0.9 and $\bar{x}$ = 7 % ±4 of the Se loss

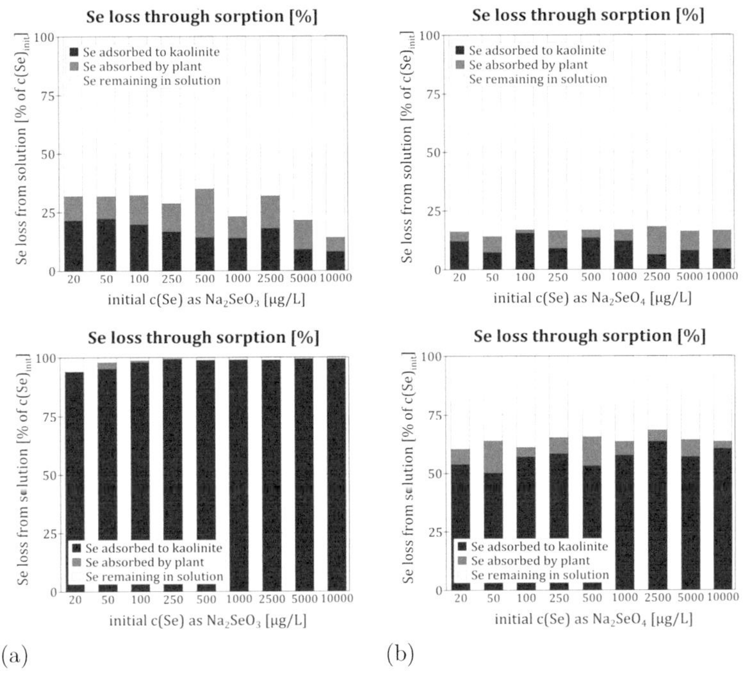

Figure 5.12: Loss of Se concentration from solution through adsorption onto kaolinite (top) and goethite (bottom)
a: addition of Se as Na_2SeO_3
b: addition of Se as Na_2SeO_4

for selenite and selenate, respectively. In Exp. B1, selenite in solution was nearly completely (96 - 97 %) adsorbed onto goethite, while Se-loss from solution was due to selenate sorption onto goethite (43 - 36 %).

Adsorption did not only affect Se concentrations in the nutrient solution. As shown in Figure 5.13, concentrations of N as nitrate, S as sulfate and P as phosphate were influenced as well; however, there was no visible correlation of nutrient adsorption with Se concentration. Kaolinite was observed to adsorb phosphate with a loss from solution of $\bar{x} = 55$ % ± 11 and $\bar{x} = 65$ % ± 7 in the selenite and selenate experiments, respectively, but no measurable amounts of nitrate ($\bar{x} = 0$ % ± 0 and $\bar{x} = 0$ % ± 1) or sulfate ($\bar{x} = 0$ % ± 0 and $\bar{x} = 0$ % ± 1). This means that Se speciation had an influence on phosphate adsorption onto kaolinite in the presence of selenite (Figure 5.13a), lowering phosphate adsorption compared to the presence of selenate (Figure 5.13b). As shown in Figure 5.13, loss of nitrate and sulfate from these solutions was due entirely to plant uptake and not kaolinite. Plant uptake, therefore, led to a loss of nitrate of $\bar{x} = 36$ % ± 20 and $\bar{x} = 30$ % ± 7 and a loss of sulfate of $\bar{x} = 11$ % ± 4 and $\bar{x} = 9$ % ± 4 in the selenite and selenate experiments, respectively.

Phosphate uptake by plants led to loss from solution of $\bar{x} = 20$ % ± 9 and $\bar{x} = 18$ % ± 5 in the selenite and selenate experiments, respectively. Being within each other's standard deviation, this allows the conclusion that the Se speciation had no significant influence on nitrate, sulfate or phosphate uptake, although it must be noted that uptake in the presence of selenite was slightly lower in all three cases.

Goethite was observed to adsorb 100 % of all phosphate independently of the Se speciation present (Figure 5.13 (bottom)), leaving at the most trace concentrations of phosphate for plant uptake. Adsorption of nitrate and sulfate was sparse, with nitrate loss from solution at $\bar{x} = 1$ % ± 2 and $\bar{x} = 4$ % ± 5 and sulfate loss from solution at $\bar{x} = 3$ % ± 3 and $\bar{x} = 7$ % ± 4 for selenite and selenate experiments, respectively. Due to the small adsorption amount, possible Se speciation influences were not significant and sulfate uptake into plants could not be measured in the presence of goethite independent of the Se speciation. Loss of nitrate from solution due to plant uptake was $\bar{x} = 28$ % ± 5 and $\bar{x} = 33$ % ± 14 in the selenite and selenate experiments, respectively, and also showed no relation to Se speciation.

When analysing the pH-values, initial values for nutrient solution with selenite (Figure 5.14a) increased with additional Se from 4.9 - 6.3

($\bar{x}$ = 5.4 ±0.4) and 4.9 - 6.0 ($\bar{x}$ = 5.1 ±0.4) for initial solutions of the kaolinite and goethite experiment, respectively – as already shown in Chapter 4.5. On the other hand, initial values for selenate solutions (Figure 5.14b) remained constant between 4.7 and 5.1 ($\bar{x}$ = 4.9 ±0.1) and between 4.5 and 4.9 ($\bar{x}$ = 4.7 ±0.1) for initial solutions of the kaolinite and goethite experiment, respectively.

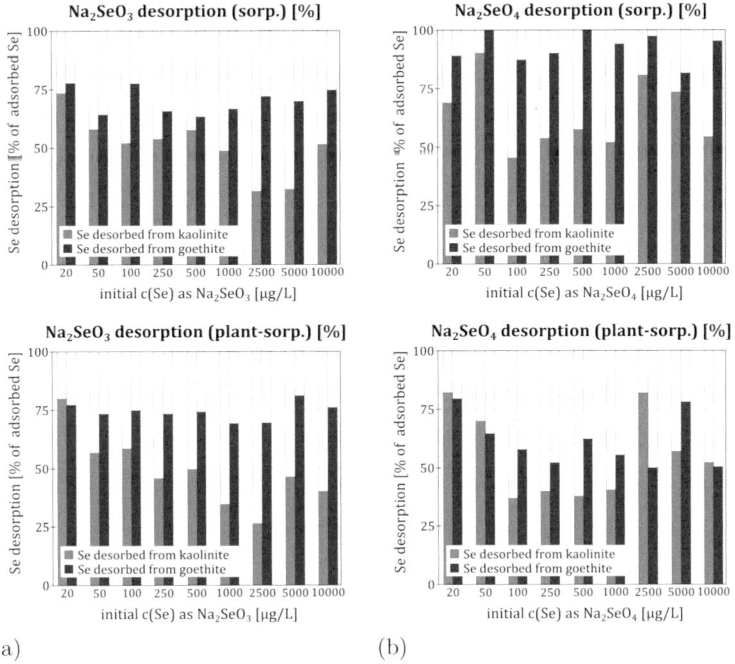

Figure 5.13: Loss of NO_3^-, SO_4^{2-} and PO_4^{3-} concentration from solution through uptake by plants (colored bar with grey-structured filling) and substrate adsorption (colored bar) onto kaolinite (top) and goethite (bottom); plant uptake calculated from the difference between plantless reference bottle experiments and rice plant-containing plant-box experiments.
a: addition of Se as Na_2SeO_3
b: addition of Se as Na_2SeO_4

The levelling effect of mineral adsorption already described in Exp. B1 and B2 (Chapter 4.5) was prominent for kaolinite, but less pronounced for goethite. Kaolinite sorption increased the pH-value to $\bar{x} = 6.7$ % ± 0.1 for both selenite and selenate adsorption, respectively, indepedent of initial solution pH. With the additional presence of plants, pH increased to $\bar{x} = 7.2$ % ± 0.1 (with one outlier at 5.8) and $\bar{x} = 7.4$ % ± 0.1 in the selenite and selenate experiments, respectively. Goethite sorption led to higher pH-values than kaolinite with $\bar{x} = 8.2$ % ± 0.3 and $\bar{x} = 7.1$ % ± 0.3 for selenite and selenate adsorption, respectively. The presence of plants levelled the pH to $\bar{x} = 7.6$ % ± 0.3 and $\bar{x} = 7.4$ % ± 0.4 in the selenite and selenate experiments, respectively.

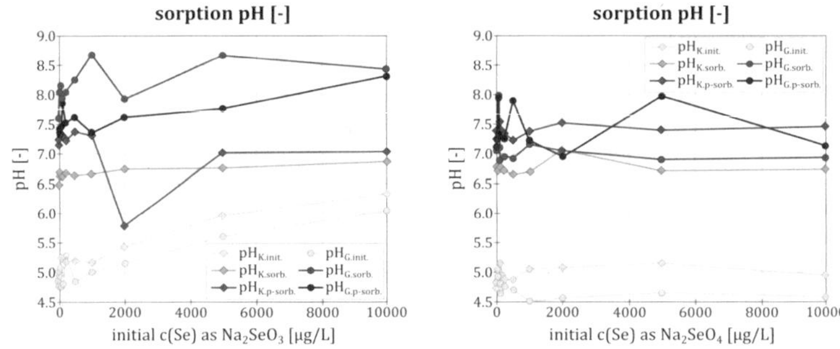

(a) (b)

Figure 5.14: pH-values before (init.) and after Se sorption (sorb.) and sorption coupled with plant-uptake (p-sorb.) onto kaolinite (K) and goethite (G)
a: as Na_2SeO_4
b: as Na_2SeO_3

5.3.5 Desorption properties

Desorption of previously adsorbed Se in Exp. C (Figure 5.15) was generally very inhomogenous, but without a recognizable Se concentration-dependent trend. As with Exp. B3, kaolinite desorption did not change the pH of the desorption solution (pH = 8.0). Selenite desorption recovery from kaolinite (Figure 5.15a) was between 32 and 74 % ($\bar{x}$ = 51 % ±13) for the pure adsorption in the glass reference bottles, and between 26 and 80 % ($\bar{x}$ = 49 % ±16) for the plant-box experiments. Selenate desorption from kaolinite (Figure 5.15b) was between 45 and 81 % ($\bar{x}$ = 62 % ±17) for the pure adsorption in the glass reference bottles, and between 37 and 82 % ($\bar{x}$ = 55 % ±19) for the plant-box experiments.

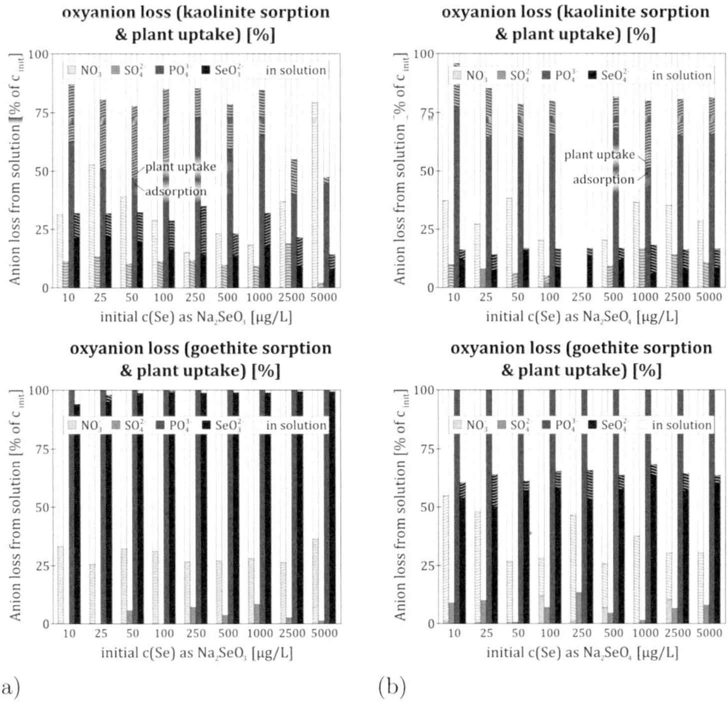

Figure 5.15: Se desorption profiles from kaolinite and goethite for reference bottle experiments (top) and plant-box experiments (bottom) using KH_2PO_4 as percentages of previously adsorbed Se.
a: desorption of selenite
b: desorption of selenate

This means that Se was less homogenously recoverable after adsorption from a nutrient solution than from pure solutions, and that desorption from kaolinite was twice as homogenous when plants were present during adsorption.

Similarly to results in Exp. B3, goethite desorption changed the pH of the desorption solution from 8.0 to 9.5. Selenite desorption recovery from goethite (Figure 5.15a) was between 63 and 78 % ($\bar{x} = 70$ % ± 6) for the pure adsorption in the glass reference bottles, and between 69 and 81 % ($\bar{x} = 74$ % ± 4) for the plant-box experiments. Selenate desorption from goethite (Figure 5.15b) was between 82 and 100 % ($\bar{x} = 95$ % ± 10) for the pure adsorption in the glass reference bottles, and between 50 and 79 % ($\bar{x} = 61$ % ± 11) for the plant-box experiments.

In general, therefore, desorption from goethite had a higher Se recovery (6 - 33 % higher) than desorption from kaolinite but overall inhomogeneity of recovery was very high (SD = 6 - 19 %) compared to Exp. B3 (SD = 3 - 8 %). Previous adsorption in the presence of plants led to slightly greater inhomogeneity of Se recovery (SD = 1 - 3 %).

5.3.6 SEM results

SEM analysis revealed no visible surface alteration due to Se sorption for either goethite or kaolinite in the presence of selenite or selenate. This was attributed to the low Se concentrations used in the experiments.

5.3.7 XANES results

Figure 5.16a shows that the measured references of selenite, selenate and SeMet species were easily distinguishable from one another. The absorption edge peaks were at 12.661 keV for SeMet, 12.664 keV for selenite and 12.667 keV for selenate. Peaks were distinct for each species;

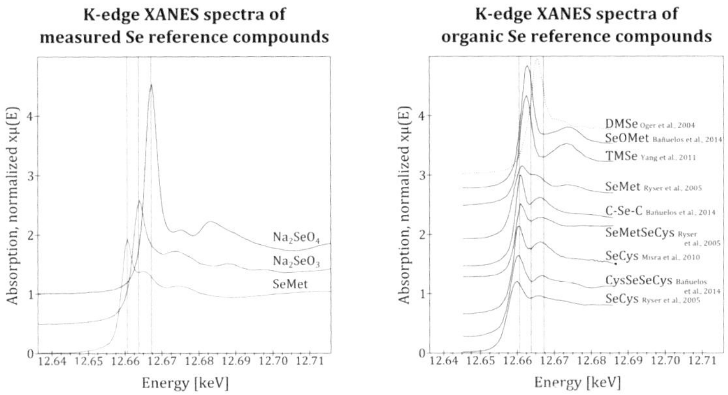

(a) (b)

Figure 5.16: XANES of reference materials, with lines in green, red and blue indicating the absorption edges of the measured standards,
a: measured references selenomethionine (SeMet, absorption edge peak: 12.661 keV), selenite (12.664 keV) and selenate (12.667 keV)
b: spectra digitized from publications for orientation and linear fitting

selenate showed the most pronounced peak (ca. 2.5 times the normed edge height), while selenite's peak was less pronounced (ca. 1 times the normed edge height), and SeMet's peak was the least prounounced (ca. 0.5 times the normed edge height). Within the XANES region ($\pm$30 eV), the ascending peak slopes were similar, but the tailing slopes showed distinct oscillations with maxima at 12.665 and 12.675 keV for SeMet, 12.673 and 12.688 keV for selenite and 12.675 and 12.683 keV for selenate.

Because there are more organic Se compounds present in a plant than the Se references measured in this study, supplementary reference data (Figure 5.16b) was aquired from various publications using a digitizing software (Engauge Digitizer, version 5.1). The range of organic Se spans peaks at 12.660 keV for SeCys [Ryser et al., 2005], 12.661 for CysSeSeCys [Bañuelos et al. 2015], SeCys [Misra, 2010] and SeMetSeCys [Ryser et

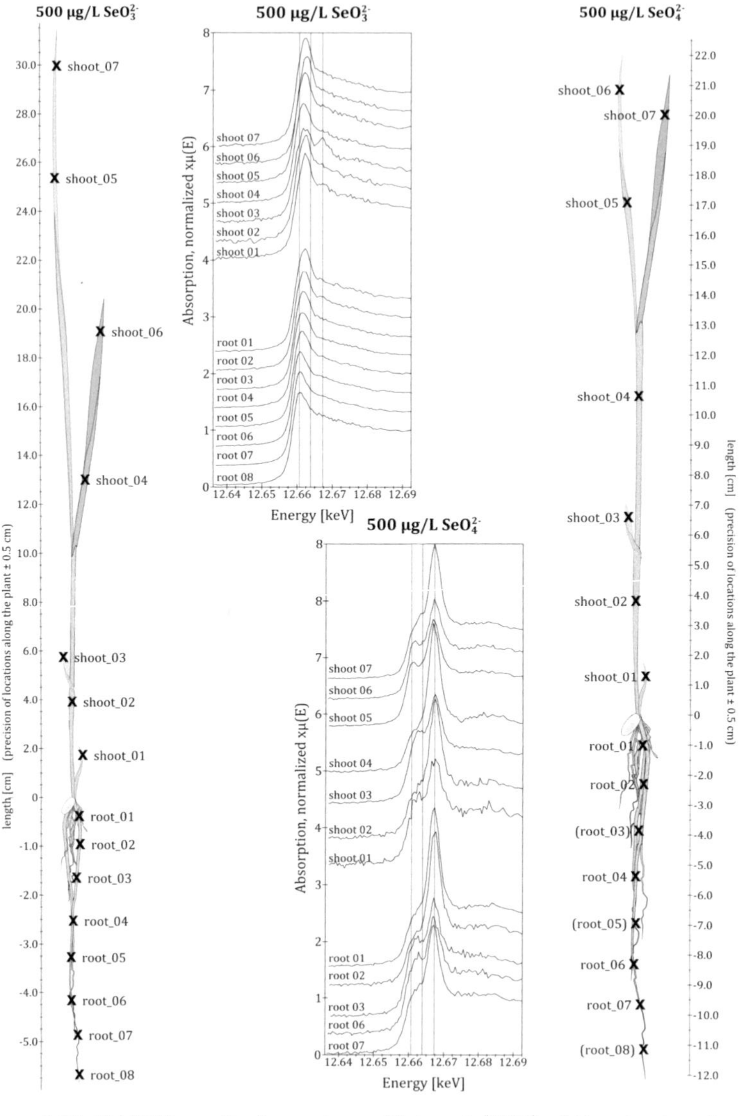

Figure 5.17: XANES results for regions of interest (ROI) of the shoot and root of a dried rice plant treated with 500 µg/L Se as Na_2SeO_3 (left) or Na_2SeO_4 (right); green, red & blue indicating peak lines for selenomethionine (12.661 keV), selenite (12.664 keV) and selenate (12.667 keV), respectively.

al., 2005], 12.662 keV for SeMet [Ryser et al., 2005] and 12.663 keV for TMSe [Yang et al., 2011] and SeOMet [Bañuelos et al. 2015] as well as 12.666 keV for DMSe [Oger et al., 2004]. The measured SeMet spectrum (Figure 5.16a) matched the organic Se mean referred to as Se-C-Se [Bañuelos et al. 2015], indicating that a distinction between SeMet and SeCys as well as combinations thereof would be difficult to make. After a linear combination fitting survey of 4 measured plant sample scans of selenite and selenate with all combinations of measured and acquired references, all subsequent linear fittings were conducted with the following 5 references as all the others appeared negligible: SeMet (measured), C-Se-C [Bañuelos et al. 2015], SeOMet [Bañuelos et al. 2015], selenite (measured) and selenate (measured). Results of SeMet, C-Se-C and SeOMet were combined and labelled "organic Se".

Figures 5.17, 5.18, 5.19 show that plants treated with selenite and selenate exhibited differences in the species found in XANES spectra of their respective plant tissues. When adding selenite to nutrient solution, peak energy of the XANES spectra generally suggested organic Se to be dominant within the plant (Figure 5.17, 5.18, 5.19). Furthermore, a depth-dependent trend was visible for the peak-energy. The absorption peak, though retaining its shape, shifted by 10 eV toward a higher energy the closer the scans were to the caryopse – from 12.661 keV at -6.0 cm (root 08) to 12.662 keV at -0.5 cm below the surface (root 01) when adding 500 µg/L Se (Figure 5.17). A much smaller peak shift of no more than 5 eV was observed with the addition of 2000 µg/L Se, while none at all was visible with the addition of 10,000 µg/L Se. Linear combination fitting revealed that only a small portion of selenite was found in the root; 0 - 27 % ($\bar{x} = 15$ % ±9), 0 - 11 % ($\bar{x} = 3$ % ±4) and 0 - 3 % ($\bar{x} = 0$ % ±1) for the additions of 500, 2000 and 10,000 µg/L Se, respectively. The proportion of organic Se, the main species found in

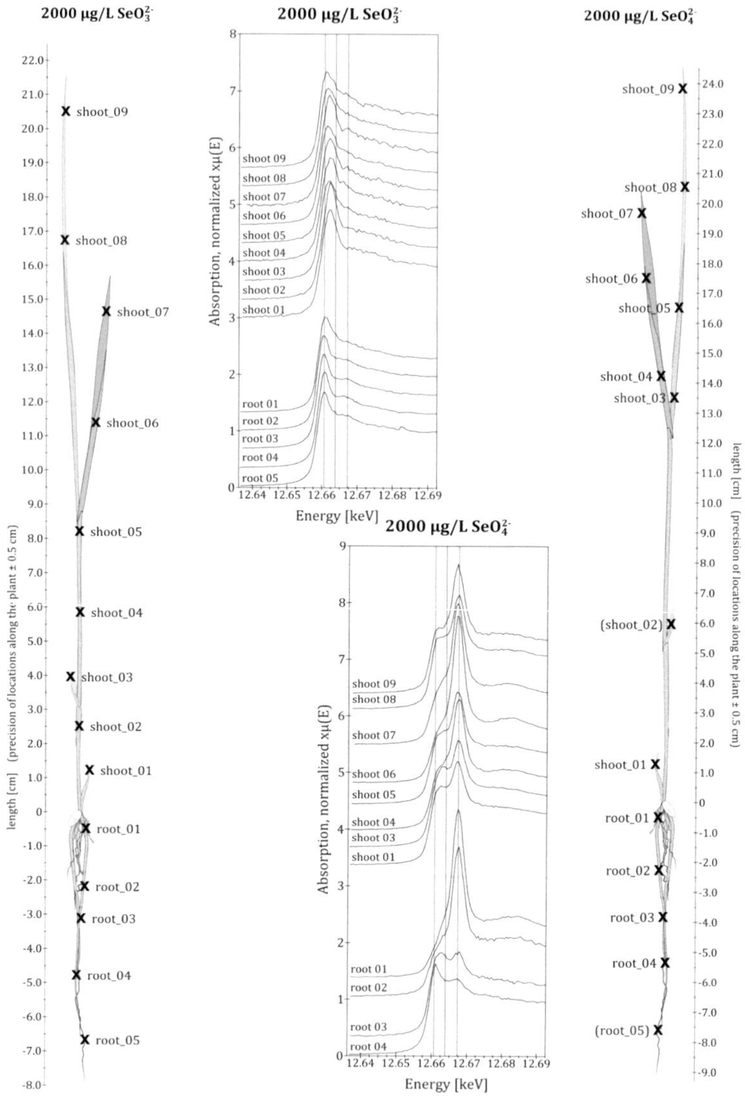

Figure 5.18: XANES results for regions of interest (ROI) of the shoot and root of a dried rice plant treated with 2000 µg/L Se as Na_2SeO_3 (left) or Na_2SeO_4 (right); green, red & blue indicating peak lines for selenomethionine (12.661 keV), selenite (12.664 keV) and selenate (12.667 keV), respectively

the root, increased toward lower root regions (root 01 - root 08; Figure 5.17, 5.18, 5.19), but also with added Se concentration as well: 73 - 100 % (between 0 and -6 cm below surface; $\bar{x}$ = 85 % ±9), 89 - 97 % (between 0 and -4.5 cm below surface; $\bar{x}$ = 95 % ±4) and 95 - 100 % (between 0 and -3 cm below surface; $\bar{x}$ = 99 % ±2) of SeMet with the additions of 500, 2000 and 10,000 µg/L Se, respectively. The proportion of selenate was calculated to be only 1 - 3 % for isolated samples and, therefore, negligible. This corresponds very well with the peak positions and shapes that correlate with SeMet to a large degree.

In the shoots, organic Se proportion was lower than in the roots with the addition of Se as selenite and speciation did not differ significantly between tissue height above ground. Increasing in proportion to added Se concentration, organic Se comprised up to 50 - 74 % ($\bar{x}$ = 65 % ±8), 63 - 76 % ($\bar{x}$ = 74 % ±11) and 74 - 81 % ($\bar{x}$ = 78 % ±3) of Se in the shoot tissue with the addition of 500, 2000 and 10,000 µg/L Se, respectively. The remaining 26 - 50 % ($\bar{x}$ = 35 % ±8), 24 - 37 % ($\bar{x}$ = 27 % ±11) and 19 - 26 % ($\bar{x}$ = 22 % ±3) of plant-Se was comprised of selenite. Shoot 09 (2000 µg/L Se added, Figure 5.18) was an exception, showing 100 % of its Se to be organic Se, while no selenite was found. No selenate was found in the shoots when selenite was added to the nutrient solution. This corresponds very well with the peak positions and shapes that appear to be a mixture of SeMet and selenite and shift toward the SeMet position and shape as Se concentrations are increased.

When treating the plants with selenate, Se speciation in the roots differed compared to treatment with selenite. Absorption spectra of the roots changed from containing more organic Se to containing more selenate between lower root scans and scans closer to the caryopse; particularly as Se concentration increased from 500 (Figure 5.17) to 2000 (Figure 5.18) and then to 10,000 (Figure 5.19) µg/L Se. As with selen-

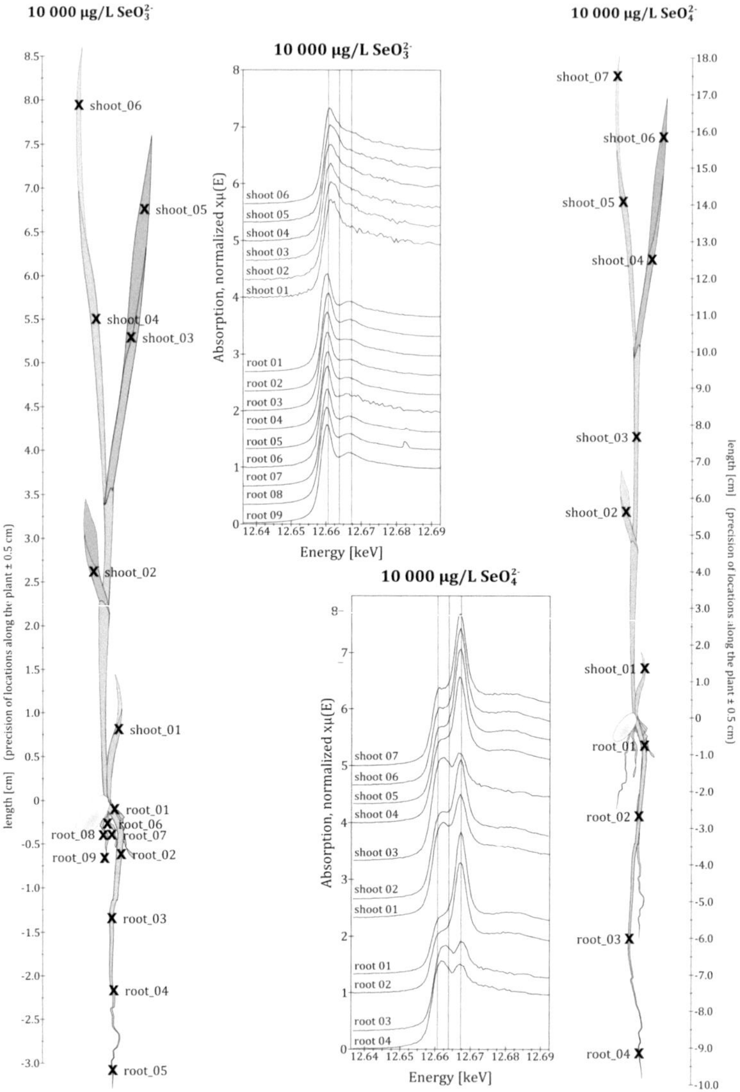

Figure 5.19: XANES results for regions of interest (ROI) of the shoot and root of a dried rice plant treated with 10,000 µg/L Se as Na_2SeO_3 (left) or Na_2SeO_4 (right); green, red & blue indicating peak lines for selenomethionine (12.661 keV), selenite (12.664 keV) and selenate (12.667 keV), respectively

ite treatment, the proportion of organic Se increased the further below surface the root sample was taken. When adding 500 µg/L Se as selenate (Figure 5.17), organic Se comprised 31 % of plant-Se for root 01 at -0.5 cm below the surface and increased to 50 % for root 06 at -8.0 cm below the surface. Correspondingly, selenate, the second major Se species in the plant-root, decreased with root depth, from 63 % for root 01 to 32 % for root 06. The calculated proportion of selenite, however, increased with depth, similar to organic Se from 6 % for root 01 to 17 % for root 07. Mean values of organic Se proportion in the roots increased with additional Se from $\bar{x} = 42$ % ±10 to $\bar{x} = 48$ % ±30 and $\bar{x} = 54$ % ±7 for the additions of 500, 2000 and 10,000 µg/L Se, respectively. Similarly to the addition of 500 µg/L Se as selenite, increased Se concentrations led to the same trends of increasing organic Se with root depth, decreasing selenate with root and increasing selenite with root depth, although this became more pronounced when increasing the added Se concentration. With the addition of 2000 µg/L Se as selenate (Figure 5.18), root-Se between root 01 at -0.5 cm and root 05 at -8.0 cm showed an increase of organic Se from 14 % to 80 % ($\bar{x} = 48$ % ±30) and a corresponding decrease in selenate from 71 % to 5 % ($\bar{x} = 36$ % ±34). Selenite made up between 10 and 27 % ($\bar{x} = 17$ % ±7) of plant-Se. With the addition of 10,000 µg/L Se as selenate (Figure 5.19), root-Se between root 01 at -0.5 cm and root 04 at -9.0 cm showed an increase of organic Se from 43 % to 68 % ($\bar{x} = 54$ % ±11), an increase of selenite from 6 % to 23 % ($\bar{x} = 17$ % ±12) and a corresponding decrease in selenate from 51 % to 9 % ($\bar{x} = 29$ % ±22).

Selenium speciation in the shoot with the addition of Se as selenate exhibited greater heterogeneity than in roots or in shoots and roots for the addition of selenite. However, there was no significant spatial trend to explain this heterogeneity. Scans show a dominance in selenate with visibly smaller peaks at the SeMet energy, which is reflected in the linear combination fitting, as well. Addition of 500 µg/L Se led to 27 - 57 % SeMet ($\bar{x} = 38$ % ±10), 6 - 28 % selenite ($\bar{x} = 14$ % ±8) and 31 - 67 % selenate ($\bar{x} = 48$ % ±16). Addition of 2000 µg/L Se led to 25 - 56 % SeMet ($\bar{x} = 45$ % ±11), 10 - 22 % selenite ($\bar{x} = 18$ % ±5) and 22 - 65 % selenate ($\bar{x} = 31$ % ±13). Addition of 10,000 µg/L Se led to 43 - 68 % SeMet ($\bar{x} = 54$ % ±11), 6 - 31 % selenite ($\bar{x} = 17$ % ±12) and 8 - 51 % selenate ($\bar{x} = 29$ % ±22). Heterogeneity decreased with increased Se concentration, and, as previously noted, average organic Se proportion increased with increased Se concentration as well, while proportions of selenite and selenate varied.

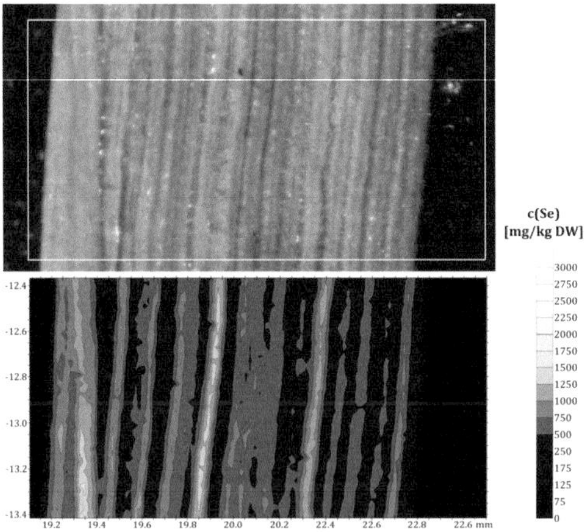

Figure 5.20: Selenium fluorescence mapping of air-dried shoot tissue – rice plant treated with 10,000 µg/L Se as selenate in hydroponic culture; photo (top) taken under the binocular

5.3.8 Results of high-resolution Se mappings using synchrotron based µXRF

The µXRF results exhibited a very distinct pattern of Se distribution, which followed the plant's vascular bundles (Figure 5.20). This was observed independently from the addition of Se as selenite or selenate. Concentrations of Se, which could be determined semi-quantitatively, were up to 2000 mg/kg DW and selectively reached 3000 µg/L in these vascular bundles of the rice plant treated with 10,000 µg/L Se (Figure 5.22). In between these vascular bundles, Se concentrations were below 500 mg/kg DW, but not below 50 mg/kg DW.

For plants treated with 2000 µg/L Se as selenite, Se content in the shoots was much lower (Figure 5.21, top). Again, concentrations were highest in the vascular bundles, but only amounted to 45 mg/kg DW Se at the most and the lowest Se concentration within the plant tissue was 10 mg/kg DW.

When the rice plant was treated with 10,000 µg/L Se as selenite (Figure 5.21, bottom), fluorescence showed vascular bundles in the shoots with Se hot spots of up to 1750 mg/kg DW Se. However, overall Se content was closer to 1000 mg/kg DW. Root cross-sections also showed the highest Se concentrations to be in the middle of the root (up to 3000 mg/kg DW) where vascular bundles converge into the central cylinder, with Se decreasing toward the outer edge of the cross-section to below 500 mg/kg DW (Figure 5.21).

In some places, although slightly deformed by the pressure of the razor blade used to produce the root cross-section of the selenite-treated plant(Figure 5.21), single vascular bundles stood out with high Se concentrations of more than 2000 µg/kg DW. Selenium content appeared to be highest for root cross-sections further below the surface with

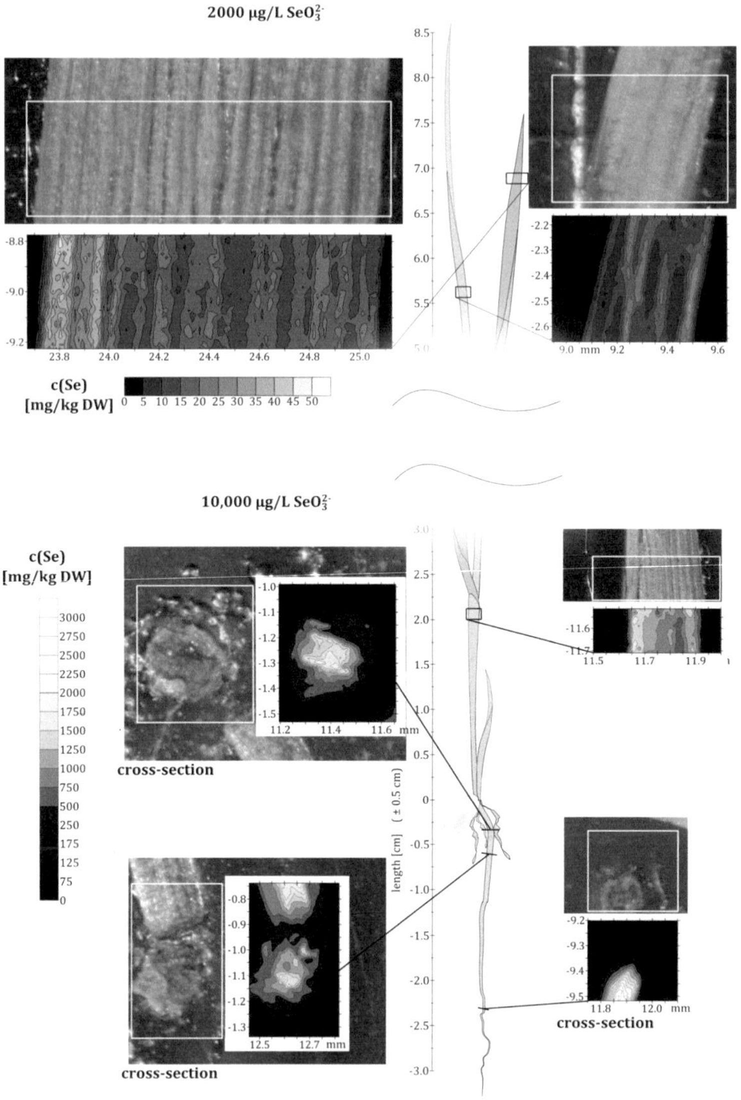

Figure 5.21: Selenium fluorescence mappings of shoot and root tissue of a dried rice plant treated with 2000 µg/L Se (top) and 10,000 µg/L Se (bottom) as Na_2SeO_3; photos taken with a binocular

3000 mg/kg DW at -2.5 cm, compared to 2000 mg/kg DW at -0.5 cm below the surface.

When the rice plant was treated with 10,000 µg/L Se as selenate, concentrations in vascular bundles of the leaves reached up to 3000 mg/kg DW Se in tissue of 15 cm growth height. Selenium content in the highest-measured leaf (4th leaf, 17.5 cm growth height) still reached 1750 mg/kg DW. Leaf tips, such as the coleoptile or the tip of the second leaf, however, showed lower Se concentrations of 500 - 1000 mg/kg DW. Root Se concentrations for the selenate-treated plants were lower than in their selenite counterparts and reached 1250 mg/kg DW at the most. It is interesting to note that root hairs, although large enough to be resolved by the X-ray beam of 11 x 16 µm if they contained Se, show no trace of Se concentration.

5.4 Discussion

As it has already been thoroughly discussed in Chapter 3.6, selenate accumulation predominantly in shoots and selenite accumulation predominantly in roots will not be discussed deeply here. Likewise, adsorption and desorption properties of kaolinite and goethite with respect to selenite and selenate as well as the nutrient oxy-anions nitrate, sulfate and phosphate have already been discussed in Chapter 4.6.

5.4.1 Selenium toxicity in plants under optimal conditions

As argued in Chapter 3.6, the availability of nutrients and germination in a Se-free environment enables the rice plant to deal with much higher Se concentrations than if grown directly in Se without nutrient supplementation. However, additions of 5000 and 10,000 µg/L Se lead

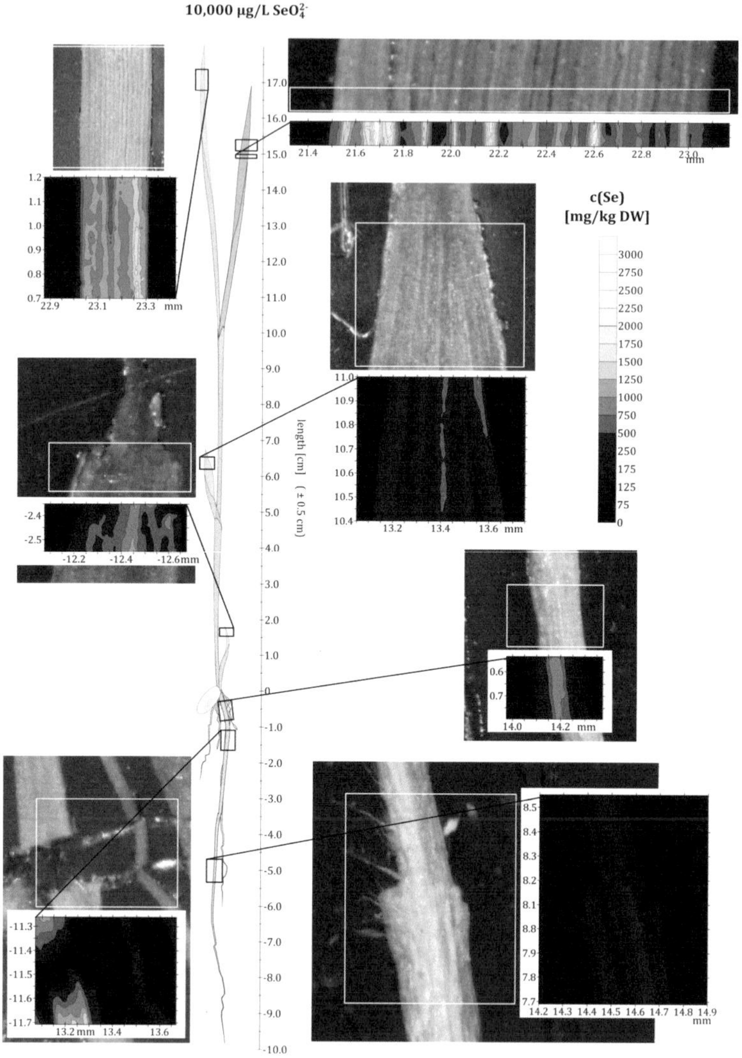

Figure 5.22: Selenium fluorescence mappings of shoot and root tissue of a dried rice plant treated with 10,000 µg/L Se as Na_2SeO_4; photos taken with a binocular

to toxicity symptoms similar to those on the agar plants despite Se-free germination and the supplement of nutrients. If the set-up in the agar experiments can be considered the maximum of adverse effects on the rice plant concerning nutrients and Se-respite during germination, these experiments constitute the minimum of adverse effects with regard to those two factors. Therefore, 2000 µg/L Se addition as selenite is the highest tolerance limit for 2 - 3 week-old rice plants under optimal conditions. Since selenate shows no adverse effect on the rice plants in these concentrations, selenite is considered more toxic than selenate.

The 2000 mg/kg DW of Se considered to lead to 10 % less crop yield in rice [Läuchli, 1993] must, therefore, refer to selenate addition, since selenite shows severe toxicity already at concentrations below 100 mg/kg DW in shoots and 500 mg/kg DW in roots. Therefore, for rice plants, selenite can be considered 8 - 12 times more toxic than selenate when considering the entire plant Se concentration, which corresponds very well with present literature, in which a similar difference in toxicity has been described [Smith & Watkinson, 1984,Läuchli, 1993]. This is particularly important because much rice is cultivated in wet soil and stagnant water, therefore making selenite the more prevalent Se species [Khan & Hell, 2014].

5.4.2 Substrate influence on Se-uptake by rice

As to be expected due to results of the previous experiments, kaolinite adsorption influenced plant-Se uptake slightly, while goethite adsorption had a major influence. Kaolinite's limited sorption capacity reduced selenite availability by 16 % and selenate availability by 10 %. This largely remained within error margins of plant uptake for previous experiments of the same cultivar and is therefore not considered significant. Therefore, soils containing kaolinite offer little protection against toxicity of high Se concentrations and are also unsuitable as a Se reservoir for Se-biofortification of plants.

Selenite and selenate adsorption to goethite, on the other hand, was significant. With 98 % Se-loss due to selenite adsorption and 57 % Se-loss due to selenate adsorption, plant-uptake of Se is strongly inhibited by goethite. Plant-uptake reflected this Se-loss. Although plant roots had physical contact with the Se-containing goethite, there is no evidence of increased uptake beyond that of pure Se content in the nutrient solution, since uptake was not elevated compared to the uptake without the presence of Se-rich sorption material. Most likely, plants which are not under stress to obtain nutrients will not desorb Se from mineral surfaces by root exudates. The fact that Se can strongly adsorb onto iron oxides and hydroxydes has been noted in many publications [Balistrieri & Chao, 1987, Hayes et al., 1987, Balistrieri & Chao, 1990, Chan et al., 2009, Das et al., 2013]. What is noteworthy as well, however, is the wide pH range for which this is the case. Due to large quantities adsorbed to the soil, a sudden change in redox and pH conditions can re-submerge these previously adsorbed quantities of Se and make them bioavailable. This is not unrealistic for rice agriculture, as flooded conditions during early growth stages alternate with dry conditions around harvest time and this leads to frequent redox changes in the soil. As shown in the desorption experiments, phosphate can easily displace adsorbed Se, even from goethite. So, not only is a sudden change of redox conditions important to monitor in such cases, but also the agricultural application of large quantities of phosphate-bearing fertilizer. Since iron oxides will preferentially bind phosphate, not only would plants be left without phosphate, but also exposed to the possibly toxic concentrations of Se.

Another interesting observation is that adsorption onto kaolinite differed compared to the pure sorption experiments, in which there was no nutrient solution, but rather a KCl background electrolyte of similar ionic strength. The presence of further nutrient ions, especially phosphate and sulfate, diminished kaolinite adsorption capacity by a factor of 2 - 4 for selenite and by a factor of 4 - 7 for selenate. Since plant germination and root growth requires up to 5 - 7 days for roots to reach the nutrient solution, this effect of kaolinite adsorption interference must be a product of the additional nutrients. It also explains why the effect is $7/4^{ths}$ more pronounced for selenate than for selenite, as this is the ratio of molar concentrations of sulfate (700 µmol) and phosphate (400 µmol). Therefore, selenite adsorbed onto kaolinite is believed to be hindered by competing phosphate ions, while selenate adsorption is hindered by competing sulfate adsorption.

5.4.3 Se distribution within the plant

As argued previously in the literature [Sun et al., 2010, Bañuelos et al. 2015], XANES spectra of some organic species are difficult to distinguish. Moreover, different authors have measured slightly different peak energies for SeCys and SeMet, which confirms that measurement results for these compounds, in particular, are not as conservative as for example measurement results for selenite or selenate, which do not vary. In a study using a different measuring technique (HPLC-ICP-MS), species of SeMet, MeSeCys, and SeCys were discernable and a linear correlation between these species was formulated for tissue samples of fruit-bearing rice plants [Sun et al., 2010]. What XANES spectra and µXRF mappings in our study were able to show in contrast, however, is the spatial progression of species proportions along the length of the root, in particular. As shown in this study, the proportion of organic Se within the root is lowest near the caryopse and increases toward lower parts of the root for both selenite and selenate. As Se is known to be rapidly converted into organic Se when taken up as selenite [Terry et al., 2000, Khan & Hell, 2014], finding a large portion of organic Se in the roots is to be expected [Terry et al., 2000, Khan & Hell, 2014, Eiche, 2015]. Finding higher concentrations of selenite in higher parts of the plant might indicate that not all selenite was transformed in the root, but rather was transported into higher regions of the plant before being transformed into organic Se there. Formation of SeCys, which precedes formation of SeMet, is believed to take place in the chloroplasts, where methylation can render it a non-protein (SeMetSeCys), protecting against Se incorporation into enzymes without sacrificing Cys [Zayed et al., 1998, Terry et al., 2000, Finley, 2005]. An explanation for why the organic Se proportion is higher in the root than in the shoot in general and for Se addition as selenate, in particular, might be that tranformation into organic Se appears to be either more effective in the roots, or that Se is stored as organic Se in the roots as a non-protein, such as MeSeCys [Duc et al., 2003], while it is volatilized from the shoots. As methylation of Se might inhibit its incorporation into proteins, thereby providing a method to avoid toxicity [LeDuc et al., 2004], it is feasable that higher Se concentrations lead to a higher proportion of SeMet in the plant, as observed with the additions of 500, 2000 and 10,000 µg/L Se.

It is known that selenate reduction, which also takes place in the chloroplasts, is the rate-limiting step of selenate assimilation [Terry et al., 2000]. Therefore, it is not surprising to find a large portion of Se as selenate in the shoots when adding Se as selenate. The question arises why so much selenite is also present, since selenite was shown to be rapidly converted into organic Se. Although there is a selenate reduction step to selenite [Terry et al., 2000, Finley, 2005], it is still an unanswered question why it isn't synthesized more quickly into organic Se, or if this is perhaps a different species not accounted for by the linear combination fitting. There have been similar findings in wheat, with selenite found in the shoots by linear combination fitting of XANES data [Eiche, 2015].

One possibility, of course, is beam-induced damage and therefore, reduction of selenate into selenite [Wang et al. 2013]. However, well aware of this danger, measurements were kept very short and no more than 2 scans were taken from the exact same spot and the reproducibility of such great quantitites of selenite due to sudden beam-induced damage when none was previously seen is highly unlikely. It is more likely that either selenite is really transported as selenite to higher parts of the plant, which would not agree with other currently published data [Läuchli, 1993, Zayed et al., 1998, Terry et al., 2000, Sun et al., 2010], or that some vital Se reference is missing in the linear fitting. The insecurities connected with the linear fitting coupled with a probable lack of reference material is currently considered the most plausible explanation for this finding. There are multiple possible Se compounds that include S which could be part of this explanation, since their XANES spectra show peaks at multiple energies between organic species and selenite [Pickering et al., 1999]. More detailed investigations are needed to determine these spatial-speciation differences.

5.4.4 Conclusions for the biogeochemical mass balance model

In combination with results of both other sets of experiments (Chapter 3.6, 4.6), these results are incorporated into the biogeochemical model in Chapter 6. Model-relevant conclusions from this part of the study not already made in previous chapters are:

1. Se-uptake for concentrations greater than 2500 µ/L Se (i.e. 5000 and 10,000 µg/L Se) can be added;
2. selenite and selenate-dependent partitioning into organic Se, selenite and selenate can be mathematically formulated for shoots and roots for a range of Se concentrations;
3. additionally, root depth profiles of this partitioning can be described dependent on Se speciation and concentration;
4. kaolinite and goethite sorption in the presence of a nutrient solution can be compared to Se adsorption in the presence of single-ion competition;
5. direct influence of substrate adsorption on bioavailable Se for plant uptake can be modelled.

6 A biogeochemical model of Se transfer in the Critical Zone

In order to connect the experimental results of this study to formulate a coherent model, all key findings were transferred into mathematical terms. These mathematical arguments of this bottom-up approach do not necessarily describe processes mono-causally, but were intended to provide a good approximation for quantities of Se transfer between reservoirs in the Critical Zone.

6.1 Selenium transfer between soil solution and plant

For this part of the model, data were used in which Se was taken up from nutrient solution, including experiments A3 (Chapter 3.3.3) and C (Chapter 6.5c). Since growing rice without nutrients is highly unlikely in a practical setting, the results of experiment A2 (Chapter 3.3.2) and A4 (Chapter 3.3.4) are not included, as those experiments were conducted on nutrient-free agar instead of nutrient solution. The following assumptions were made for the model, resulting from Exp. A3 (Chapter 3.3.3) and Exp. C (Chapter 6.5c):

- differences between selenite and selenate uptake into plants made separate modelling of each species necessary
- differences between root and shoot partitioning within the rice plants required separate modelling of these plant tissues
- therefore, speciation differences also made separate modelling of root and shoot necessary
- in addition to the mean root speciation, local root speciation could be modelled as a function of depth below the surface

6.1.1 Data input

For the model of resulting Se-content in the plant, all data of measured (Exp. A3) or sorption-corrected Se (Exp. C) as selenite or selenate in the initial solution was correlated with plant uptakes into shoots and roots. To provide a large data-set for this modelling, data of all 3 replicates as well as the mean were plotted. Mathematical arguments for the fitting were achieved with Excel (Microsoft, version 14.0.7147.5001) and pre-setted trend lines were fitted. Best fits were chosen by comparing R^2.

Since the uptake of 10,000 µg/L Se was only covered by two data points, it was not included in the model. Because the range of Se concentration was large enough to produce two very different fits when either including or omitting data on the uptake of 5000 µg/L Se, both types of fits were used in the model. Additionally, smaller concentration ranges (0 - 50 µg/L Se, 50 - 500 µg/L Se and 500 - 5000 µg/L Se) were modelled as well, using only the mean values of Exp. A3 and Exp. C. These were included optionally in the model.

Table 6.1: Root depth dependence of linear combination fitting results; selenite addition required only percentages of organic Se data, selenate addition required data for percentages of organic Se and selenate

root depth [cm]	**root-Se org. [%]** c(Se) as SeO_3^{2-} [µg/L]			**root-Se org. [%]** c(Se) as SeO_4^{2-} [µg/L]			**root selenate [%]** c(Se) as SeO_4^{2-} [µg/L]		
	500	2000	10,000	500	2000	10,000	500	2000	10,000
0	73		100		14	43	6	10	6
-1	76			31			6	15	8
-2	79	89	100	32	32	48			
-3	84	97	100						
-4	85	100		50		57	17		31
-5	91	99	100	53	63	68		27	
-6	100				46		18	27	27
-7	92	94							23

For the model of Se speciation percentages within roots and shoots, mean results of the linear combination fitting of Exp. C were used. Therefore, proportions of three species were modelled: organic Se, selenite and selenate. Data on tissue speciation distribution for 500, 2000 and 10,000 µg/L Se in the initial solution allowed concentration-dependent speciation modelling. In the case of selenite addition, amounts of selenate found in the tissue were negligible and were set to 0. Only organic Se content was modelled over a Se concentration range of 500 - 10,000 µg/L and extrapolated to lower Se concentrations; selenite content in the tissue was then obtained by subtracting the organic content from 100 %. In the case of selenate addition, differences in root and shoot speciation trends meant modelling had to be adapted. Therefore, roots required modelling of organic Se and selenite, while the remaining selenate fraction was obtained by subtracting the sum of organic Se and selenite from 100 %. For the shoots, organic Se and selenate percentages were modelled, while the remaining selenite fraction was obtained by subtracting the sum of organic Se and selenate from 100 %.

Not only was concentration-dependent Se speciation modelling possible, but also depth-dependent speciation shifts in the root as well. This required a 3-step fitting. The first step was the mean root speciation dependent on Se concentration described above. In the second step, linear (y = mx + t) Se concentration-dependent deviation from the mean was determined for each root depth by plotting the share of organic Se at a depth of x cm per mean root organic Se against the added Se concentration. This required Se speciation to be known for 2 - 3 Se concentrations (500, 2000 and 10,000 µg/L Se) in the same root depth (Table 6.1). Due to the depth-dependence of this speciation, each root-depth resulted in plots with different slopes (m) and y-axis intercepts (t) in its respective mathematical argument. In the third step, these (m and t) were plotted against root depth. Therefore, depth-dependent Se speciation was modelled as the linear deviation from the Se concentration-dependent root-speciation mean with slope and y-axis intercepts dependent on root depth.

6.1.2 Fitting results

Fitting plant-Se content over the span of 0 - 5000 µg/L Se in the nutrient solution (Figure 6.1) produced polynomial functions of the 2nd degree. Equations 6.1 and 6.2 describe shoot-Se content, which were fitted with an accuracy of R^2 = 0.9604 and R^2 = 0.9859 with the addition of 0 - 5000 µg/L Se as selenite and selenate, respectively.

$$c_{P.sh} = -2.0 \cdot 10^{-5} c_{SeO_3}^2 + 0.1013 c_{SeO_3} - 0.5859 \tag{6.1}$$

$$c_{P.sh} = -5.0 \cdot 10^{-6} c_{SeO_4}^2 + 0.147 c_{SeO_4} - 0.8039 \tag{6.2}$$

with

$c_{P.sh}$ as the mean Se shoot content [mg Se/kg DW]

c_{SeO_3} as the initial bioavailable solution selenite concentration [µg/L Se]

c_{SeO_4} as the initial bioavailable solution selenate concentration [µg/L Se]

Root-Se content was modelled as shown in Equations 6.3 and 6.4, which were fitted with an accuracy of $R^2 = 0.971$ and $R^2 = 0.9744$ with the addition of 0 - 5000 µg/L Se as selenite and selenate, respectively.

$$c_{P.r} = -2.0 \cdot 10^{-5} c_{SeO_3}^2 + 0.177 c_{SeO_3} + 5.3887 \tag{6.3}$$

$$c_{P.r} = -1.0 \cdot 10^{-6} c_{SeO_4}^2 + 0.0471 c_{SeO_4} + 0.0659 \tag{6.4}$$

with

$c_{P.r}$	as the mean Se root content [mg Se/kg DW]
c_{SeO_3}	as the initial bioavailable solution selenite concentration [µg/L Se]
c_{SeO_4}	as the initial bioavailable solution selenate concentration [µg/L Se]

Mirroring the experimental results, the polynomial describing plant-Se due to selenite treatment shows a decrease in uptake at concentrations higher than 2500 µg/L Se (Figure 6.1a). However, when fitting plant-Se content over the span of 0 - 2500 µg/L Se rather than 0 - 5000 µg/L, thereby omitting this change of Se-uptake pattern, the accuracy of the fitting was not improved, neither for shoot-Se ($R^2 = 0.958$ vs. $R^2 = 0.9604$) nor for root-Se ($R^2 = 0.9660$ vs. $R^2 = 0.9710$).

Plant-Se uptake due to selenate treatment (Figure 6.1b) could be described by a straight line rather than a polynomial. Omitting data for the addition of 5000 µg/L Se and using only datasets of experiment C, both for shoot-Se ($R^2 = 0.9966$ vs. $R^2 = 0.9859$) and root-Se ($R^2 = 0.9976$ vs. $R^2 = 0.9744$) improved fitting. Equations 6.5 and 6.6 describe Se shoot content with an accuracy of $R^2 = 0.958$ and $R^2 = 0.9966$ with the addition of 0 - 2500 µg/L Se as selenite and selenate, respectively.

$$c_{P.sh} = -1.0 \cdot 10^{-5} c_{SeO_3}^2 + 0.0974 c_{SeO_3} - 0.1734 \tag{6.5}$$

$$c_{P.sh} = 0.133 c_{SeO_4} + 0.7698 \qquad (6.6)$$

with

$c_{P.sh}$ as the mean Se shoot content [mg Se/kg DW]

c_{SeO_3} as the initial bioavailable solution selenite concentration [µg/L Se]

c_{SeO_4} as the initial bioavailable solution selenate concentration [µg/L Se]

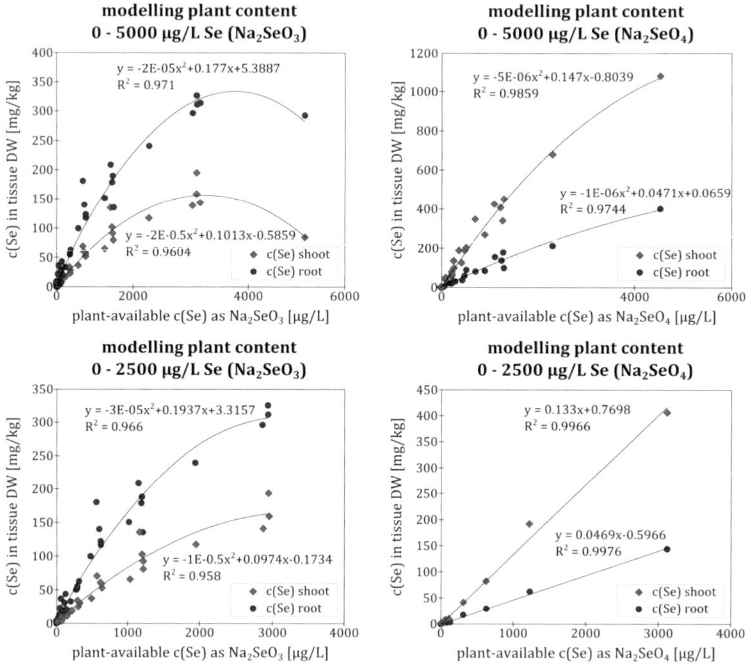

(a) (b)

Figure 6.1: Excel fitting of experimental data on plant uptake in a range of 0 - 5000 µg/L Se in the nutrient solution (top) with a polynomial and in a range of 0 - 2500 µg/L Se in the nutrient solution (bottom) with a polynomial or straight line
a: plant uptake of Se as Na_2SeO_3 into shoots and roots
b: plant uptake of Se as Na_2SeO_4 into shoots and roots

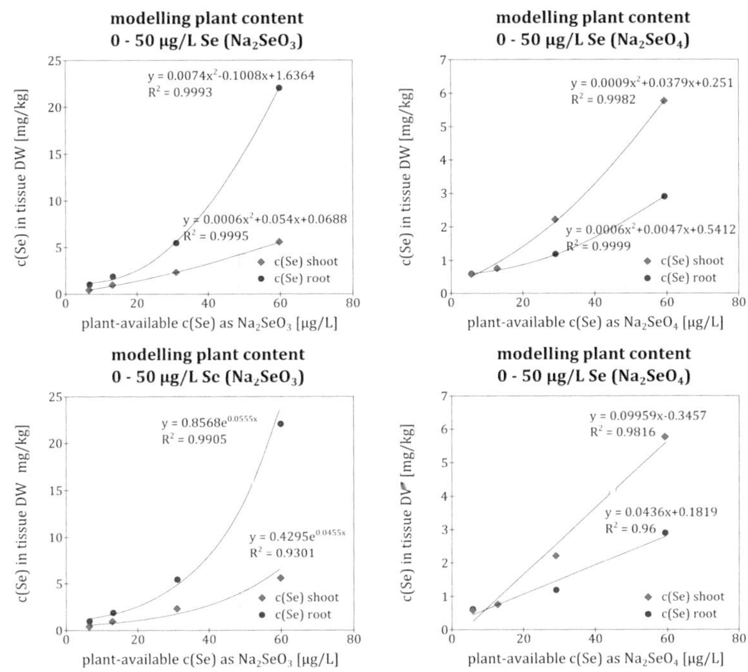

(a) (b)

Figure 6.2: Excel fitting of experimental data on plant uptake in a range of 0 - 50 µg/L Se in the nutrient solution
a: plant uptake of Se as Na_2SeO_3 into shoots and roots fitted with a polynomial (top) or exponential curve (bottom)
b: plant uptake of Se as Na_2SeO_4 into shoots and roots fitted with a polynomial (top) or straight line (bottom)

Equations 6.7 and 6.8 describe Se root content with an accuracy of $R^2 = 0.966$ and $R^2 = 0.9976$ with the addition of 0 - 2500 µg/L Se as selenite and selenate, respectively.

$$c_{P.r} = -3.0 \cdot 10^{-5} c_{SeO_3}^2 + 0.1937 c_{SeO_3} + 3.3157 \quad (6.7)$$

$$c_{P.r} = 0.0469 c_{SeO_4} - 0.5966 \quad (6.8)$$

with

$c_{P.r}$ as the mean Se root content [mg Se/kg DW]

c_{SeO_3} as the initial bioavailable solution selenite concentration [µg/L Se]

c_{SeO_4} as the initial bioavailable solution selenate concentration [µg/L Se]

Overall, fitting quality was limited most by scattering (1s error) of experimental data, which was discussed in Chapter 3.5.3.

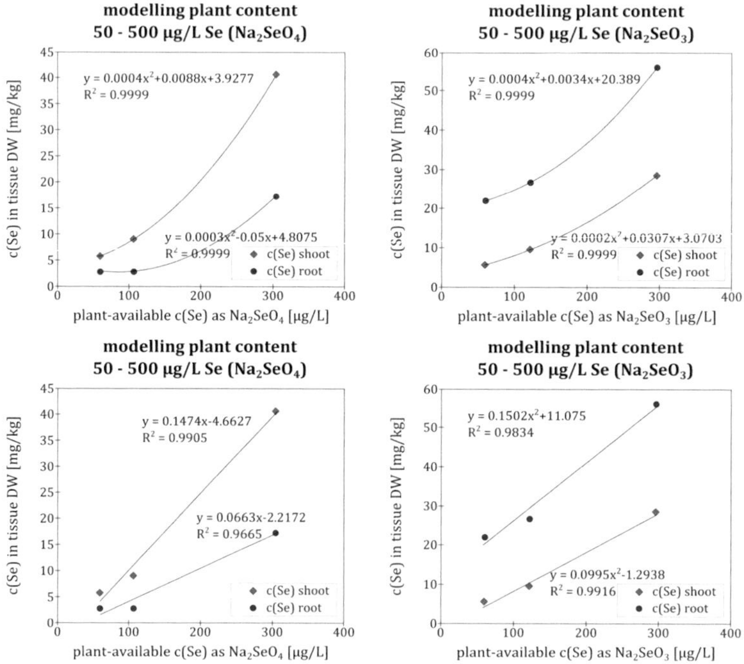

(a) (b)

Figure 6.3: Excel fitting of experimental data on plant uptake in a range of 50 - 500 µg/L Se in the nutrient solution
a: plant uptake of Se as Na_2SeO_3 into shoots and roots fitted with a polynomial (top) or power function (bottom)
b: plant uptake of Se as Na_2SeO_4 into shoots and roots fitted with a polynomial (top) or straight line (bottom)

Modelling of smaller ranges of Se treatment yielded better accuracy and worked best for polynomial functions of the 2nd degree. Addition of 0 - 50 µg/L Se (Figure 6.2) was modelled with an accuracy of $R^2 = 0.9995$ and $R^2 = 0.9993$ in shoots and roots for selenite addition and with an accuracy of $R^2 = 0.9982$ and $R^2 = 0.9999$ in shoots and roots for selenate addition, respectively.

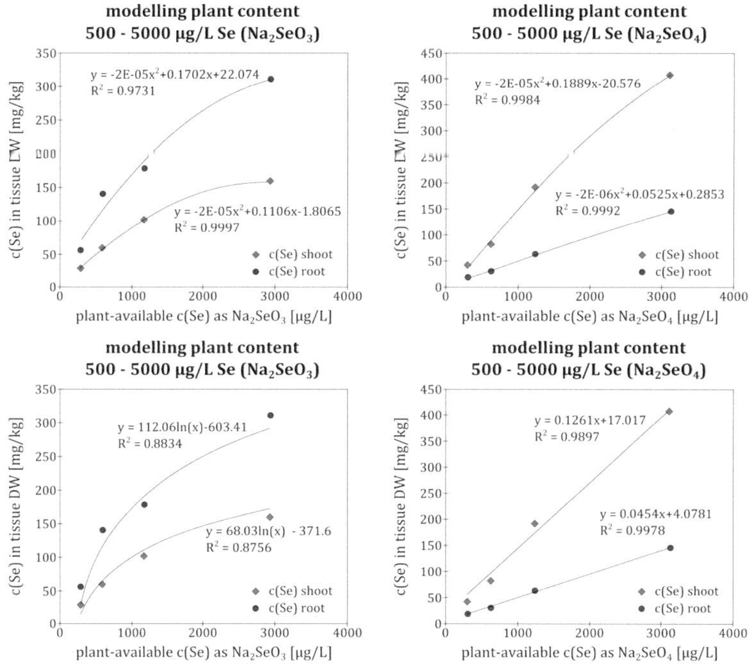

(a) (b)

Figure 6.4: Excel fitting of experimental data on plant uptake in a range of 500 - 5000 µg/L Se in the nutrient solution
a: plant uptake of Se as Na_2SeO_3 into shoots and roots fitted with a polynomial (top) or logarithmic function (bottom)
b: plant uptake of Se as Na_2SeO_4 into shoots and roots fitted with a polynomial (top) or straight line (bottom)

Addition of 50 - 500 µg/L (Figure 6.3) Se was modelled with an accuracy of $R^2 = 0.9999$ in shoots and roots for selenate and selenite addition, respectively. Addition of 500 - 5000 µg/L Se (Figure 6.4) was modelled with an accuracy of $R^2 = 0.9997$ and $R^2 = 0.9731$ in shoots and roots for selenite addition and with an accuracy of $R^2 = 0.9984$ and $R^2 = 0.9992$ in shoots and roots for selenate addition, respectively.

While plant-Se in shoots and roots resulting from the addition of Se as selenite was alternatively modelled using an exponential fitting (0 - 50 µg/L Se, Figure 6.2), a squared function (50 - 500 µg/L Se, Figure 6.3) and a logarithmic function (500 - 5000 µg/L Se, Figure 6.4), the resulting plant-Se in shoot and root after addition of Se in the form of selenate was alternatively fitted linearly for each concentration range. However, in general, polynomial fitting of the 2nd degree provided the best accuracy. In the overall model, a mean of the fittings for plant-Se in ranges of Se addition of 0 - 5000 µg/L Se and 0 - 2500 µg/L Se was primarily used. However, results for all fittings of the smaller Se ranges are given as well for purposes of comparison.

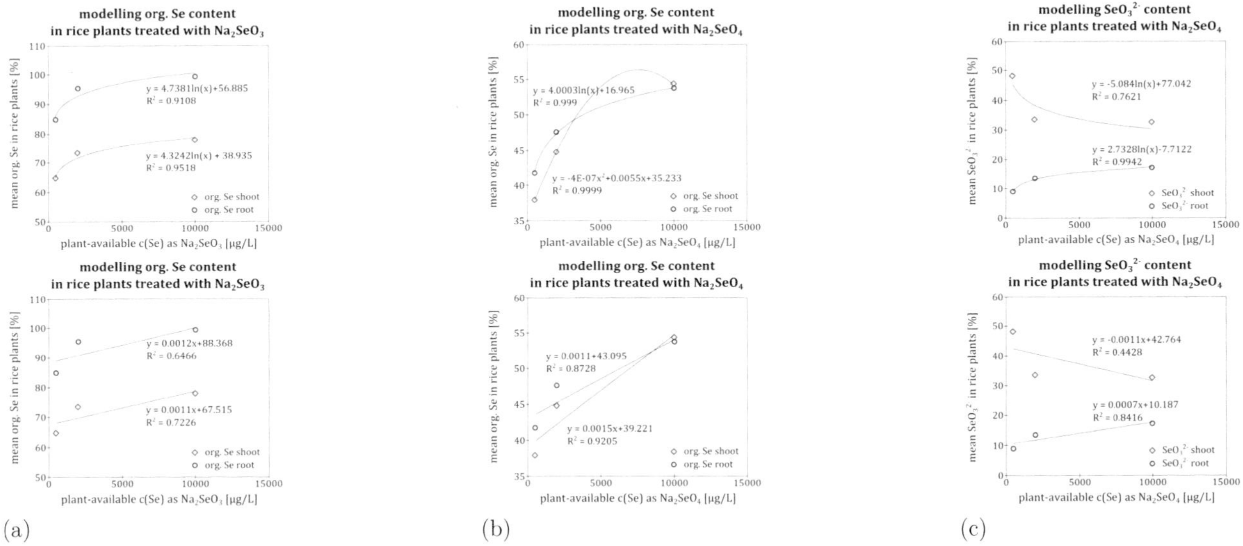

Figure 6.5: Excel fitting of plant tissue speciation according to XANES data with plant uptake for nutrient solution concentrations of 500, 2000 and 10,000 µg/L Se
a: mean percentage of organic Se in shoots and roots when Se is added as Na_2SeO_3 fitted with logarithmic function (top) or straight line (bottom)
b: mean percentage of organic Se in shoots and roots when Se is added as Na_2SeO_4 fitted with logarithmic function (top) or straight line (bottom)
c: mean percentage of selenite in shoots and roots when Se is added as Na_2SeO_4 fitted with logarithmic function (top) or straight line (bottom)

Fitting the mean XANES results for Se speciation after Se uptake resulted in a much lower accuracy than plant-Se content modelling. As shown in Figure 6.5, a logarithmic fitting showed greater accuracy than linear fittings; organic Se in shoots was fitted with an accuracy of $R^2 = 0.9518$ and $R^2 = 0.9999$ with the addition of selenite and selenate, respectively, while organic Se in roots was fitted with an accuracy of $R^2 = 0.9108$ and $R^2 = 0.9990$ for the addition of selenite and selenate, respectively, As shown in Figure 6.5b, polynomial fitting for organic Se in shoots of rice plants treated with selenate demonstrated the best fitting. For all other data sets, the 3rd degree polynomial fitting – although exhibiting the highest accuracy in describing the three data points – resulted in a curve progression highly unlikely to represent speciation and was, therefore, discarded. With XANES data sets providing speciation information in the plant tissue only for 500, 2000 and 10,000 µg/L Se, however, any attempted extrapolation of this information on speciation in plant tissue for lower Se concentrations, therefore, has an unquantifiable error.

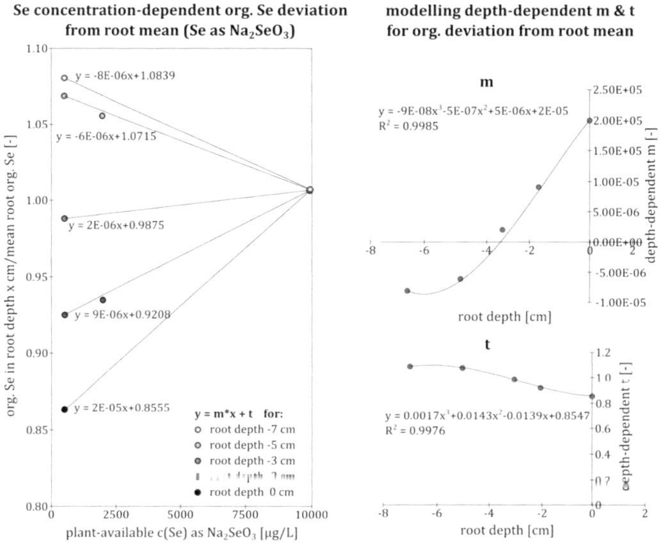

Figure 6.6: Root speciation modelling for additions of Se as Na_2SeO_3;
left: fitting of the deviation of local organic Se content compared to the root mean organic Se content for 5 root depths as measured for 500 and 10,000 µg/L Se added as selenite.
right: root depth-correlated fitting of parameters (m, t) describing the organic Se speciation pattern of all local deviations from root mean

As shown in Figure 6.6, the pattern for localized root speciation along the length of the root when the plant was treated with selenite was fitted linearly with 2 - 3 data points per Se concentration. Parameters of depth dependence for this linear concentration-dependent fitting had a high accuracy for both m ($R^2 = 0.9985$) and t ($R^2 = 0.9976$). For the model of the organic percentage of depth-dependent root-Se, parameters m and t were substituted by these fittings, making distribution of organic Se - and, therefore, the distribution of selenite as well – dependent on three parameters: Selenium concentration in the solution, the previously calculated mean root organic Se (which is dependent only on the Se concentration in the solution), and the local root depth below the surface.

The resulting modelled functions (Equations 6.9, 6.10, 6.11, 6.12) for the percentage of organic Se in any root depth between 0 and -7 cm below the surface for a known selenite concentration in the nutrient solution were therefore:

$$p_{org.Se_r} = P_{\bar{x}Se_r} \cdot (m \cdot c_{SeO_3} + t) \tag{6.9}$$

and

$$P_{\bar{x}Se_r} = \frac{(4.7381 \ln c_{SeO_3} + 56.885) + ((0.0012 c_{SeO_3} + 88.368))}{2} \tag{6.10}$$

and

$$m = -9.0 \cdot 10^{-7} d_r^3 - 5.0 \cdot 10^{-7} d_r^2 + 5.0 \cdot 10^{-5} d_r + 2.0 \cdot 10^{-4} \tag{6.11}$$

and

$$t = 0.0017 d_r^3 + 0.0143 d_r^2 - 0.0139 d_r + 0.83357 \tag{6.12}$$

with

$p_{org.Se_r}$	as the local percentage of organic Se in the root [%]
$P_{\bar{x}Se_r}$	as the mean percentage of organic Se in the root [%]
c_{SeO_3}	as the initial bioavailable solution selenite concentration [µg/L Se]
d_r	as the root depth below the surface [cm]

The local percentage of organic Se in the root when Se is added as selenate was calculated, fitted and modelled in similar fashion. However, Se concentration dependence was fitted with the greatest accuracy when using a power function (Figure 6.7). Resulting function parameters needed to model depth-dependence, were, therefore, m_1, m_2 and t.

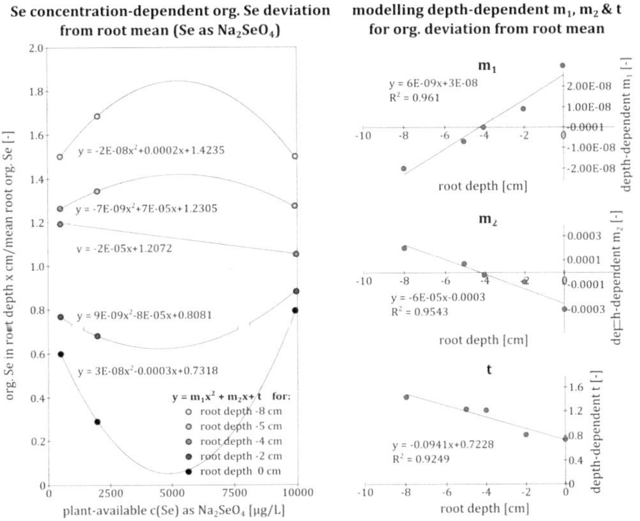

Figure 6.7: Root speciation modelling for additions of Se as Na_2SeO_4;
left: fitting of the deviation of local organic Se content compared to the root mean organic Se content for 5 root depths as measured for 500, 2000 and 10,000 µg/L Se added as selenate.
right: root depth-correlated fitting of parameters (m, t) describing the organic Se speciation pattern of all local deviations from root mean

These were fitted with linear functions yielding accuracy measures of $R^2 = 0.9610$, 0.9543 and 0.9249 for m_1, m_2 and t, respectively. Equations 6.13, 6.19, 6.15, 6.16 and 6.17 describe the local organic Se distribution in the root with the addition of selenate.

$$p_{org.Se_r} = P_{\bar{x}Se_r} \cdot (m_1 c_{SeO_4}^2 + m_2 c_{SeO_4} + t) \tag{6.13}$$

and

$$P_{\bar{x}Se_r} = \frac{(4.0003 \ln c_{SeO_4} + 16.965) + (0.0011 c_{SeO_4} + 43.095)}{2} \tag{6.14}$$

and

$$m_1 = 6.0 \cdot 10^{-8} d_r + 3.0 \cdot 10^{-7} \tag{6.15}$$

and

$$m_2 = -6.0 \cdot 10^{-4} d_r + 3.0 \cdot 10^{-3} \tag{6.16}$$

and

$$t = -0.0941 d_r + 0.7228 \tag{6.17}$$

with

$p_{org.Se_r}$	as the local percentage of organic Se in the root [%]
$P_{\bar{x}Se_r}$	as the mean percentage of organic Se in the root [%]
c_{SeO_4}	as the initial bioavailable solution selenate concentration [µg/L Se]
d_r	as the root depth below the surface [cm]

This fitting approach was also applied to the local percentage of selenite in the root when Se was added as selenate. This needed to be calculated separately, as well, because selenate treatment yielded three species in the root, rather than two, as for selenite treatment. As previously found with organic Se, Se concentration dependence was fitted with the greatest accuracy when using a power function (Figure 6.8) and the resulting function parameters needed to model depth-dependence were m_1, m_2 and t.

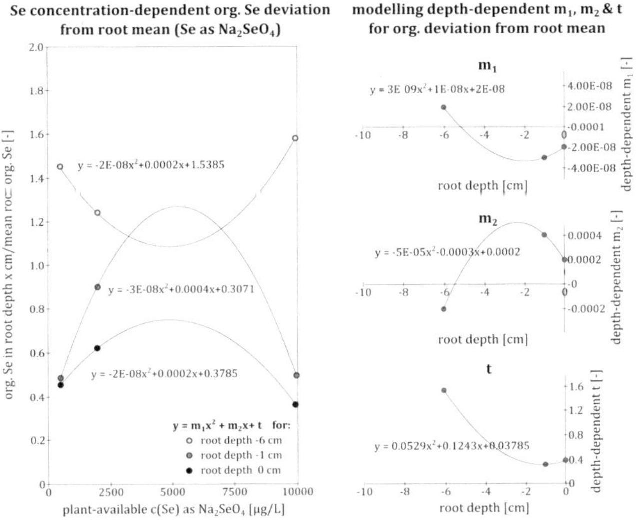

Figure 6.8: Root speciation modelling for additions of Se as Na_2SeO_4;
left: fitting of the deviation of local selenite content compared to the root mean selenite content for 5 root depths as measured for 500, 2000 and 10,000 µg/L Se added as selenite.
right: root depth-correlated fitting of parameters (m, t) describing the selenite speciation pattern of all local deviations from root mean

However, these could not be fitted with linear functions and were also described by power functions instead. Equations 6.13, 6.19, 6.15, 6.16 and 6.17 describe the local organic Se distribution in the root with the addition of selenate.

$$p_{SeO_{3r}} = P_{\bar{x}_{Se_r}} \cdot (m_1 c_{SeO_4}^2 + m_2 c_{SeO_3} + t) \tag{6.18}$$

and

$$P_{\bar{x}_{Se_r}} = \frac{(2.7328 \ln c_{SeO_4} - 7.7122) + (0.0007 c_{SeO_4} + 10.187)}{2} \tag{6.19}$$

and

$$m_1 = 3.0 \cdot 10^{-8} d_r^2 + 1.0 \cdot 10^{-7} d_r - 2.0 \cdot 10^{-7} \tag{6.20}$$

and

$$m_2 = -5.0 \cdot 10^{-4} d_r^2 - 3.0 \cdot 10^{-3} d_r - 2.0 \cdot 10^{-3} \tag{6.21}$$

and

$$t = 0.0529 d_r^2 + 0.1243 d_r + 0.3785 \tag{6.22}$$

with

$p_{SeO_{3r}}$ as the local percentage of selenite in the root [%]

$P_{\bar{x}_{Se_r}}$ as the mean percentage of organic Se in the root [%]

c_{SeO_4} as the initial bioavailable solution selenate concentration [µg/L Se]

d_r as the root depth below the surface [cm]

6.1.3 Model output

With the input of the Se concentration in solution either as selenite or as selenate between 0 and 5000 µg/L, the model provides results on Se content and speciation distribution in shoots and in roots. Furthermore, the model predicts the depth profile of speciation in the roots. Figure 6.9 exemplifies model output for 35 µg/L Se in the nutrient solution, which is a concentration not covered by the experiments.

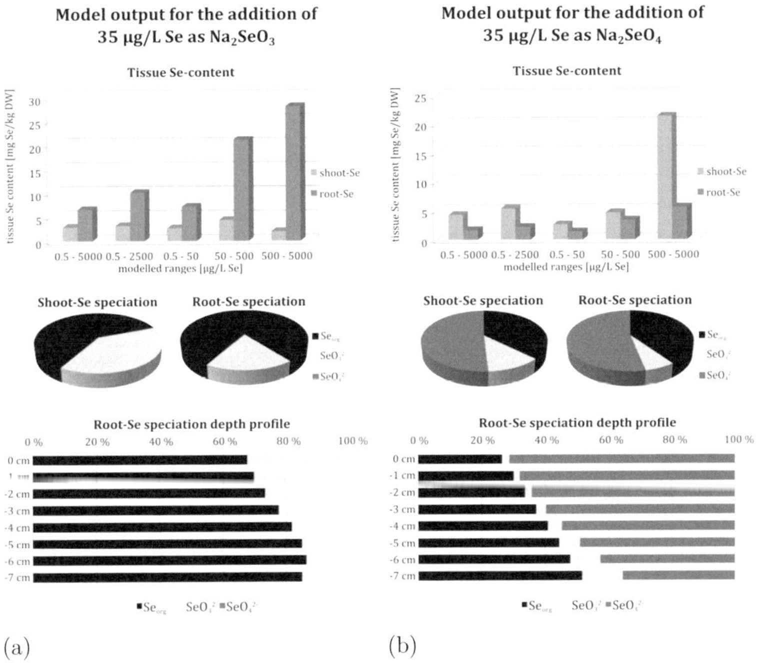

Figure 6.9: Model output for plant Se-content and speciation
a: for the addition of 35 µg/L Se as selenite
b: for the addition of 35 µg/L Se as selenate

Model output values for Se content and speciation in shoots and roots were designed to be valid for input Se concentration between 0 and 5000 µg/L Se. However, input of low Se concentrations and very high concentrations are less reliable; Se content modelling with the addition of Se as selenite, for example, yields negative results for shoot-Se below an input of 6 µg/L Se when using a mean result of Equations 6.1 and 6.5. This is not the case for the more precise modelling limited to the range of 0 - 50 µg/L Se; however, the model by default uses the overall fitting for the large concentration range.

Since this mean shoot-Se result is then used for the calculation of shoot speciation, this results in unrealistic speciation distribution and, therefore, mean-Se input needs to be adjusted to using the model more suitable for lower concentrations (Figure 6.2). Because it is so difficult to model plant-Se for the entire Se concentration range and subdivisions can be modelled with very different functions, this suggests a Se concentration-dependent change in plant Se uptake. Fitting results suggest that low Se concentrations (up to 50 µg/L Se) are taken up readily and can therefore be fitted nearly exponentially, while concentrations greater 50 µg/L can be modelled linearly with a smaller slope and concentrations greater than 1000 µg/L Se as selenite show a logarithmically describable tendency to approach an uptake maximum. Selenate concentrations between 1000 and 5000 µg/L, however, can still be described with a linear function, suggesting a load maximum at higher concentrations than the ones modelled here.

Model coherency, expressed as the standard deviation of two function results describing the same process for any given value, varies significantly across species and plant tissue; results for Se content using the models for the ranges 0 - 5000 and 0 - 2500 µg/L Se (Equations 6.2 - 6.8) are modelled with a standard deviation of 10 % or lower for selenite concentrations of 27 - 1450 µg/L Se for shoots and 220 - 3300 µg/L Se for roots and selenate concentrations of 45 - 5000 µg/L Se for shoots and 68 - 5000 µg/L Se for roots. If a standard deviation of 33 % or lower is sufficient, model results can be used for inputs of selenite concentrations of 12 - 3000 µg/L Se for shoots and 30 - 4800 µg/L Se for roots and selenate concentrations of 20 - 5000 µg/L Se for shoots and 17 - 5000 µg/L Se for roots.

Within these concentration ranges, modelled mean speciation appears plausible as well. Unfortunately, speciation was not experimentally obtained for low Se concentrations and error below 500 µg/L Se input Se concentration is not quantifiable. This means that speciation modelling on this sparse data basis is highly speculative, unlike the Se content in the plant, which is backed by a larger amount of experimental data. Nonetheless, this sort of modelling can provide estimations for comparison that might be helpful for future research. Although fitting results were best with logarithmic approaches (Figure 6.5), the progression of logarithmic fitting toward lower Se concentrations was considered unrealistic, as this would lead to near-zero percentages of organic Se compared to near-100 of selenite due to low Se in the solution. Therefore, mean results of linear and logarithmic fittings were considered more realistic, but cannot be verified.

As the error of modelling mean-Se is incalculable, modelling localized depth-dependent speciation in the root, which is dependent on this mean value, provides seemingly realistic results with an uncalculable error. However, results of negative speciation are not realistic. Therefore, input Se concentrations yielding negative speciation for either organic Se or selenite are considered to be outside of the range of valid Se concentration input. Localized root speciation profiles for depths of -1 cm to -7 cm are available with this model only for inputs of 12 - 2500 µg/L Se as selenite and 20 - 1350 µg/L Se as selenate.

6.2 Selenium transfer between soil minerals and soil solution

For this part of the model, adsorption data from Exp. C (Chapter 6.5c) were primarily used. Unfortunately, data from Exp. B1 - B3 (Chapter 4.4.2, b2, b3) could not be used for this model, as influences of nutrient components did not add up linearly and data on the competing oxy-anions were insufficient to plot ion-concentration dependencies. Furthermore, goethite could not be used for the model, since it did not reach its loading capacity in the experiments performed and maximum sorption remained unknown for this study. Therefore, no mixtures of kaolinite and goethite could be modelled, since it remained unknown how small amounts of goethite would need to be in order to show kaolinite contribution to sorption in the case of selenite. The following assumptions resulting from Exp. C (Chapter 6.5c), were made for the model:

- as shown in Exp. B1 and Exp. B3, kaolinite adsorption produced an isotherm shape, even if this was not evident from the pure data of Exp. C
- therefore, the general mathematical formula of all Se adsorption isotherms of kaolinite should be the same with only factors and constants changing, depending on experimental parameters
- selenite and selenate adsorption are modelled separately, but with the same general mathematical formula
- lower kaolinite amounts show a linear decrease of sorption capacity

6.2.1 Data input

Model input for adsorption was derived solely from data of Exp. C (Chapter 6.5c), as data of experiments performed in Exp. B1 - B3 were not conducted in a nutrient solution. However, general information on isotherm shape observed in Exp. B1 and B3 for different sorption influences determined the choice of fitting function. In addition to fitting with pre-setted trend line options of Excel (Microsoft, version 14.0.7147.5001) and respective outputs of R^2, the fitting of a Freundlich isotherm (Equation 4.4) was compared as well. Functions describing the adsorption were by default fitted to cross the y-axis at $y = 0$. As previously discussed, neither ion competition, nor goethite adsorption were integrated into the model.

6.2.2 Fitting results

As shown in Figure 6.10, from a mathematical point of view, the Freundlich isotherm described the experimental data adequately with $R^2 = 0.9907$ and 0.9801 for selenite and selenate treatment, respectively. Regardless of R^2 values, however, kaolinite was shown to have limited adsorption capacity for the Se concentrations used in Exp. B1 and B3 (Chapter 4.5.2). Therefore, the modelled Freundlich isotherms, which did not reflect this, were discarded for both selenite and selenate adsorption onto kaolinite.

Instead, adsorption of Se onto kaolinite was fitted with a polynomial of the 2nd degree, which more closely resembled the isotherm shape observed in general for kaolinite adsorption, although R^2 was only 0.9679 and 0.8735 for adsorption of selenite and selenate, respectively. Furthermore, two data points of the selenate adsorption experiment (Figure 6.10b) were discarded as outliers as well, as the polynomial calculated in their presence did not resemble an isotherm shape. Moreover, all other data points correlated very well with an isotherm shape approaching load maximum and the polynomial describing it ($R^2 = 0.9998$). Adsorption of Se was therefore modelled using Equation 6.23 for selenite and Equation 6.24 for selenate.

$$c_{sorpSeO_3} = -1.0 \cdot 10^{-7} c_{SeO_3}^2 + 0.0028 c_{SeO_3} \tag{6.23}$$

$$c_{sorpSeO_4} = -6.0 \cdot 10^{-8} c_{SeO_4}^2 + 0.0025 c_{SeO_4} \tag{6.24}$$

with

$c_{sorpSeO_3}$ as the selenite concentration load on kaolinite [mg Se/kg kaolinite]

c_{SeO_3} as the initial bioavailable solution selenite concentration [µg/L Se]

$c_{sorpSeO_4}$ as the selenate concentration load on kaolinite [mg Se/kg kaolinite]

c_{SeO_4} as the initial bioavailable solution selenate concentration [µg/L Se]

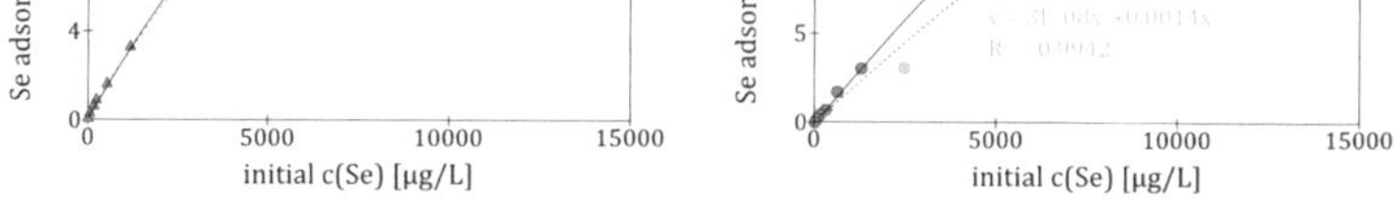

(a) (b)

Figure 6.10: Fitting of Freundlich isotherms with exponential arguments (dotted line) and polynomial to model Se adsorption onto kaolinite using isotherm-conform data (solid line) or using all data points (light grey solid line)
a: adsorption of selenite
b: adsorption of selenate

6.2.3 Model output and mathematical extrapolations

As long as adsorption of selenite and selenate is modelled by their respective isotherms dependent only on the added Se concentration, the sorption model output is described precisely by those functions. However, optional mathematically estimated parameters were included as well, as described in the following. Since these were not verified in experiments, their errors cannot be determined.

For crude estimations, the amount of kaolinite can be varied as the percentage of the experimental conditions (100 % being the 8.5 g used in the experiment with the solution unchanged at 170 mL). Although helpful for estimations, a simple percentage or factor cannot take kinetic processes into account, such as longer diffusion pathways for fewer molecules. This means that error increases, the less experimentally evaluated the kaolinite amount is. Since ion competition in Exp. B2 showed the isotherm shape to remain stable for a great number of adsorption-lowering processes, values for > 25 % of the experimental kaolinite amount are considered reasonably reliable. A similar problem is encountered at values > 100 % of the kaolinite amount used in the experiment. This is especially difficult because the maximum load capacity is unknown. Maximum load capacity calculated from sorption modelling showed a maximum Se adsorption for 14,500 µg/L Se as selenite and 21,000 µg/L Se as selenate. Although this is 3 and 4 times the amount considered to be a valid input for the model, inhibitory kinetic processes such as diffusion competition are not included and the load capacity is only extrapolated. Therefore, input values for kaolinite amounts should not exceed 150 % for the model to be considered reasonably reliable.

A second mathematically implemented parameter is that of competition between selenite and selenate. Without this implementation, adsorption of selenite and selenate were modelled separately and their combined effect was described by adding their sorption. This is reasonable as long as one of both ions is dominant in its concentration or concentrations of both ions are low. It is unreasonable, however, to expect high concentrations of each species to be adsorbed as though the other were not present, especially since kaolinite has a limited adsorption capacity and both ions adsorb at the same binding sites, albeit in a different manner (Figure 4.6).

$$c_{XsorpSeO_3} = c_{sorpSeO_3} - \left(\frac{c_{SeO_4}}{c_{SeO_4} + c_{SeO_3}} \cdot c_{sorpSeO_3} \frac{c_{sorpSeO_4}}{c_{MAXsorpSeO_4}}\right) \quad (6.25)$$

with

$c_{XsorpSeO_3}$	as selenite sorption with selenate competition [mg Se/kg kaolinite]
$c_{sorpSeO_3}$	as selenite sorption without selenate competition [mg Se/kg kaolinite]
$c_{sorpSeO_4}$	as selenate sorption without selenite competition [mg Se/kg kaolinite]
c_{SeO_3}	as the initial bioavailable solution selenite concentration [µg/L Se]
c_{SeO_4}	as the initial bioavailable solution selenate concentration [µg/L Se]
$c_{MAXsorpSeO_4}$	as the maximum selenate sorption load [mg Se/kg kaolinite]

Therefore (Equation 6.25), adsorption amounts in competition between both ions were calculated as the pure anion adsorption, from which the proportion of the percentage of the competing ion of its respective maximum load was subtracted proportional to its concentration and vice versa.

6.3 Mass balance model

Using the modelled compartments "mineral surface", "nutrient solution" and "plant", a mass balance model was created (Equation 6.26). The model assumed 100 % of the Se to be distributed into these three compartments and resulting Se content of those three compartments was given either as a percentage of total Se in the system or in µg as absolute Se amounts. To keep the model simple, the only input parameters required were the concentrations of selenite and selenate in the initial solution. Model calculations started with Se adsorption onto the mineral surface (Equation 6.27) in the presence of both competing ions (Equation 6.25). The adsorbed Se amount was then substracted from the Se in the initial solution (Equation 6.29). The resulting solution Se concentration was then used as the initial solution for subsequent Se-uptake into plant roots and shoots (Equation 6.28). This amount of Se is then substracted from the solution, yielding the final Se concentration in the solution in the mass balance (Equation 6.29).

$$m_{totalSe} = m_{Se_{sorp}} + m_{Se_{plant}} + m_{Se_{sol}} \tag{6.26}$$

and

$$\begin{aligned} m_{Se_{sorp}} = & ((c_{SeO_3} + c_{SeO_4}) - ((c_{SeO3} - \frac{m_K \cdot c_{X\,sorpSeO_3}}{V_{exp}}) + \\ & (c_{SeO4} - \frac{m_K \cdot c_{X\,sorpSeO_4}}{V_{exp}}))) \cdot V_{exp} \end{aligned} \tag{6.27}$$

and

$$m_{Se_{plant}} = \frac{N_{PB}(\frac{m_{sh}}{V_{exp}} \cdot c_{P.sh} + \frac{m_r}{V_{exp}} \cdot c_{P.r})(c_{SeO_3} + c_{SeO_4})}{(c_{SeO3} - \frac{m_K}{V_{exp}} \cdot c_{X\,sorpSeO_3}) + (c_{SeO4} - \frac{m_K}{V_{exp}} \cdot c_{X\,sorpSeO_4})} \cdot V_{exp} \tag{6.28}$$

and

$$m_{Se_{sol}} = m_{totalSe} - (m_{Se_{sorp}} + m_{Se_{plant}}) \tag{6.29}$$

with

$m_{totalSe}$	as the total Se amount in the modelled system [µg]
$m_{Se_{sorp}}$	as the Se amount adsorbed to kaolinite [µg]
$m_{Se_{plant}}$	as the Se amount taken up into the plant [µg]
$m_{Se_{sol}}$	as the Se amount remaining in solution [µg]
c_{SeO_3}	as the initial solution selenite concentration [µg/L Se]
c_{SeO_4}	as the initial solution selenate concentration [µg/L Se]
$c_{XsorpSeO_3}$	as selenite sorption with selenate competition [mg/kg]
$c_{XsorpSeO_4}$	as selenate sorption with selenate competition [mg/kg]
m_K	as the substrate amount in experiment [= 8.5 g]
V_{exp}	as the nutrient solution volume in experiment [= 0.17 L]
N_{PB}	as the mean number of plants per box [= 7.5]
$c_{P.sh}$	as the mean Se shoot content [mg Se/kg DW]
$c_{P.r}$	as the mean Se root content [mg Se/kg DW]
m_{sh}	as the mean shoot mass [= 0.01121582 g]
m_r	as the mean root mass [= 0.00621504 g]

6.4 Requirements for model extensions

This model offers a number of possibilities for extension. Criteria which must be met for this to be possible depend on the kind of extension proposed; six of which will be explored here:

1. volatilization of Se by rice plants can be added to the model using experimental measurements of rice seedlings in a comparable setting: similar growth stage, closed-box system, constant air volume and similar concentration range of Se uptake; the presence of substrate is not required; however, nutrient concentrations comparable to the nutrient solution are;

2. change in uptake behavior due to a lack of nutrients can be integrated; however, this must be coupled with adsorption studies of those same ions;

3. different plant species can be integrated as well if growth stage permits cultivation in closed plant-box systems; nutrient uptake must be monitored hereby as well;

4. goethite or other sorption substrates can be added if solution volume-to-substrate mass ratios are similar to those used in experiments and maximum load capacity is known or can be modelled; plants are not necessarily required to be part of the experimental set-up;

5. ion competition can be integrated into the model when measuring the influence on Se adsorption of multiple concentrations of multiple ions for a span of Se concentrations at a solution volume-to-substrate mass ratio comparable to the experiments carried out in this study; in this case, detailed studies in the presence and absence of rice plants are required as well, as nutrient uptake and competition affect Se uptake;

6. likewise, pH influences can be measured as well; in this case, detailed studies both in the presence and absence of rice plants are required, as nutrient uptake and competition affect Se uptake.

6.5 Evaluation and application of the mass-balance model

The model presented in this study can serve several purposes. For one, it can simply be used to interpolate results between the experimentally determined results, i.e. plant uptake calculations for 600 µg/L Se instead of the measured increments of 500 and 1000 µg/L Se or adsorption of 1800 µg/L Se onto kaolinite rather than the increments of 1000 and 2000 µg/L Se. Together with speciation information provided by the model, potential toxicity to livestock when used in fodder can be estimated, as well as the return of Se into the soil, if plants are not removed. Using data from a study on Se speciation in rice grain compared to the rest of speciation in rice [Sun et al., 2010], this model can be used to estimate grain Se content as well. The fact that different functions were required to fit the plant-Se data depending on the added Se concentration, is indicative of regulatory mechanisms of Se uptake in the plant. Even if the model cannot provide mechanistic understanding in this case, estimates for Se concentration ranges of interest can be obtained by applying the model.

While all these applications are useful, however, this model also attempted to interpolate results not obtained by the experiments, such as solutions containing a mixture of Se species or varied amounts of kaolinite. The difficulty of using such purely empirical mathematical extrapolations is that the error is uncalculable and model output must first be verified by experimental results. Therefore, this model still requires a number of validation experiments. The mean Se speciation in plant tissue needs to be verified for small Se concentrations. Moreover, more detailed data are needed to improve upon the localized root speciation.

Additional extentions to the model that can improve upon its significance would be further substrates that are frequently encountered in soils, such as iron oxides and hydroxides, other clay minerals and calcite at varying pH-values and competing nutrient anions. If experimental data was acquired with the same parameters, mixtures of minerals can be modelled to simulate a large variety of soils. An interesting further step would be to integrate organic matter as well. With the exception of microorganisms, this would constitute a complete synthetic soil. Furthermore, a greater variety of plant models would prove useful in detailed uptake modelling. Not only could additional agronomically important plants, such as wheat or corn be investigated, but also, if the experiments were to be extended over a longer time period, the fruit-bearing stage would provide insight into Se cycling throughout the entire crop cycle.

To extend the model into another compartment of the Critical Zone, trapping experiments for volatile Se species should be conducted as well. Current experiments did not focus on this aspect, as the method of volatile Se trapping was not established, yet. Therefore, the current model works under the assumption that Se not accounted for in plants or by adsorption must have remained in the solution. As previous publications have shown, however, rice is known to produce volatile Se species [Terry et al., 2000], for example, 50 - 80 % of Se added as SeMet can be volatilized. Rates indicate that selenite and selenate volatilization are lower by a factor of 40 and 20, respectively [Zayed et al., 1998].

This model provides a basis for investigations of Se transfer in the Critical Zone and can be used with a limited amount of data input. Therefore, it provides a quick overview of expected Se concentrations in kaolinite-dominated soil, solution and rice plants and is, therefore, a practical tool for first-stage assessments of Se inventory.

7 Conclusions

The objective of this study was to quantify and characterize Se transfer pathways and processes as part of the Se cycle within the Critical Zone. The strength of this integrated approach was to connect all findings by using the same experimental parameters for all aspects of this study in order to create an empirical process model of the Se biogeochemistry in the Critical Zone. This included, in particular, the substrate-to-solution ratio of sorption experiments and plant experiments as well as the ionic strength, pH-value and chemical composition of the nutrient solution in adsorption and plant uptake experiments.

By using an experimental set-up that reduces secondary influences on the Se transfer processes usually present in natural systems, such as microbiological activity, multiple adsorbing substrates or organic matter, this study was able to focus on the core transfer processes and their immediate effects. Compartmentalization of the Critical Zone proved to be a good means to explore Se transfer processes and quantification of Se transfer was possible due to the experimental set-up of closed-box systems. The two main pathways of Se transfer were investigated: the speciation- and concentration-dependent uptake of Se into rice plants and the speciation-, ion competition- and concentration-dependent Se adsorption-desorption processes between soil and nutrient solution. This study has increased knowledge and provided quantitative data on the following points:

1. Se sorption processes are an important factor controlling bioavailability and are highly dependent on mineral substrate, pH-value and nutrient ion competition. Goethite, with its large sorption capa-

cities and a great affinity for selenite adsorption, reduces selenate bioavailability to 40 % and selenite bioavailability to 0 %, while kaolinite reduces selenate bioavailability to 85 % and selenite bioavailability to 90 %. Ion competition of phosphate is strongest for both selenite and selenate adsorption, reducing Se adsorption to 15 - 50 % while sulphate reduces selenate adsorption to 15 - 75 %, but has little effect on selenite adsorption.

2. Se uptake into plants and its partitioning into different plant tissue are highly dependent on bioavailable Se concentration, Se species and oxy-anion competition in the solution as well as the plant's state of health and plant stage at the time of Se addition. Direct germination in Se and nutrient-deficiency significantly lower plant tolerance to Se and result in toxicity effects and reduced uptake at solution-Se concentrations above 250 µg/L Se as selenate and 1000 µg/L Se as selenite, while healthy plants are able to take up Se in the presence of 10,000 µg/L Se. Treatment with different Se species leads to preferential partitioning of selenite to the root (70 %) and selenate to the shoot (72 %) as well as species-dependent tissue speciation. Selenite treatment led to 85 - 100 % org. Se in roots and 64 - 80 % organic Se in shoots, the rest being selenite, and selenate treatment led to 42 - 54 % org. Se and 29 - 36 % selenate in roots and 38 - 54 % org. Se and 33 - 48 % selenate in shoots, the rest being selenite.

3. The biogeochemical process model showed plant uptake to be influenced by the sorption processes and showed inhibition of Se adsorption onto mineral surfaces by competing ions in the nutrient solution to be a non-additive process.

The mass balance model created with the data of this study can be used as a basis for further model extensions and optimizations. The model can be extended with single parameter that influence Se bioavailaility, such as pH-values, ionic strength, temperature, etc., and with additional compartments of the Critical Zone, such as the atmosphere with its Se transfer path of Se volatilization by plants. Moreover, other plants and substrates can be modelled in a similar fashion and integrated into the model as well. Other agricultural plants such as maize, beans, potatos and tomatos as well as Se accumulators such as mustard might be attractive expansions to the model due to different uptake and partitioning ratios, which is of interest depending on the plant tissue used for consumption or animal feed. In combination with additional substrates, such as ferric oxides and hydoxides, other clay minerals, carbonates and organic matter, this model might prove a useful tool to compare differences between Se bioavailability in soils comprised of these materials in various proportions without needing to experimentally investigate every iteration. Future research might benefit from the continuous development of such a model to asess optimal Se biofortification as well as Se toxicity or deficiency hazards in the field and the management thereof. All in all, Se remains an element full of challenges to science, with great potential for analytical improvements and biogeochemical pathway clarifications, toward which this study has provided one stepping stone for the improved management of the world's Se resources.

Bibliography

[Adriano, 1986] D.C. Adriano: *Chapter 12: Selenium*, 1986, in D.C. Adriano: (Ed.): *Trace Elements in Terrestrial Environments*, Springer-Verlag, New York, pp. 707-758.

[Amoroux, 2001] D. Amouroux, P.S. Liss, E. Tessier, M. Hamren-Larsson, O.F.X. Donard: *Role of oceans as biogenic sources of selenium*, 2001, Earth and Planetary Science Letters, Vol. 189, pp. 277-283.

[Anderson et al., 2007] S.P. Anderson, F. von Blanckenburg, A.F. White: *Physical and chemical controls on the Critical Zone*, 2007, Elements, Vol. 3, pp. 315-319.

[Appel et al., 2003] C. Appel, L.Q. Ma, R.D. Rhue, E. Kennelley: *Point of zero charge determination in soils and minerals via traditional methods and detection of electroacoustic mobility)*, 2003, Geoderma, Vol. 113, pp. 77-93.

[Arvy, 1993] M.P. Arvy: *Selenate and selenite uptake and translocation in bean plants (Phaseolus vulgaris)*, 1993, Journal of Experimental Botany, Vol. 44, pp. 1083-1087.

[Asher et al., 1977] C.J. Asher, G.W. Butler, P.J. Peterson: *Selenium Transport in Root Systems of Tomato*, 1977, Journal of Experimental Botany, Vol. 28, pp. 279-291.

[Bacon & Davidson, 2007] J.R. Bacon, C.M. Davidson: *Is there a future for sequential chemical extraction?*, 2007, The Analyst, Vol. 133, pp. 25-46.

[Bajaj et al., 2011] M. Bajaj, E. Eiche, T. Neumann, J. Winter, C. Gallert: *Hazardous concentrations of selenium in soil and groundwater in North-West India*, 2011, Journal of Hazardous Materials, Vol. 189, pp. 640–646.

[Balistrieri & Chao, 1987] L.S. Balistrieri, T.T. Chao: *Selenium Adsorption by Goethite*, 1987, Soil Science of America Journal, Vol. 51, pp. 1145-1151.

[Balistrieri & Chao, 1990] L.S. Balistrieri, T.T. Chao: *Adsorption of selenium by amorphous iron oxyhydroxide and manganese dioxide*, 1990, Geochimica et Cosmochimica Acta, Vol. 54, pp. 739-751.

[Bañuelos et al. 2015] G.S. Bañuelos, I. Arroyo, I.J. Pickering, S.I. Yang, J.L. Freeman: *Selenium biofortification of broccoli and carrots grown in soil amended with Se-enriched hyperaccumulator Stanleya pinnata*, 2015, Food Chemistry, Vol. 166, pp. 603-608.

[Bar-Yosef & Meek, 1987] B. Bar-Yosef, D. Meek: *Selenium Sorption By Kaolinit And Montmorillonite*, 1987, Soil Science, Vol. 144, pp. 11-19.

[Bellet al., 1992] P.F. Bell, D.R. Parker, A.L. Page: *Contrasting Selenate-sulphate Interactions in Selenium-Accumulating and Non-accumulating Plant Species*, 1992, Soil Science Society of America Journal, Vol. 56, pp. 1818-1824.

[Berzelius, 1817] J.J. Berzelius: *Lettre de M. Berzelius a M. Berthollet sur deux metaux nouveaux*, 1817, Annales de Chimie et de Physique, Vol. 7, pp. 199-207.

[Beytut et al., 2002] E. Beytut, F. Karatas, E. Beytut: *Lambs with White Muscle Disease and Selenium Content of Soil and Meadow Hay in the Region of Kars, Turkey*, 2002, The Veterinary Journal, Vol. 163, pp. 214-217.

[Birringer et al., 2002] M. Birringer, S. Pilawa, L. Flohé: *Trends in selenium biochemistry*, 2002, Natural Product Reports, Vol. 19, pp. 693-718.

[Bitterli et al., 2010] C. Bitterli, G.S. Bañuelos, R. Schulin: *Use of transfer factors to characterize uptake of selenium by plants*, 2010, Journal of Geochemical Exploration, Vol. 107, pp. 206-216.

[Boullis, 1997] B. Boullis: *Retraitement et séparation des radionucleides à vie longue*, 1997, in R. Turkey (Ed.): *Les déchets nucléaires*, Lesitions de Physique, Paris, pp. 69-91.

[Bowie et al., 1996] G.L. Bowie, J.G. Sanders, G.F. Riedel, C.C. Gilmour, D.L. Breitburg, G.A: Cutter, D.B. Porcella: *Assessing Selenium cycling and accumulation in aquatic systems*, 1996, Water Air Soil Pollution, Vol. 90 pp. 93-104.

[Brantley et al., 2007] S.L. Brantley, M.B. Goldhaber, K.V. Ragnarsdottir: *Crossing Disciplines and Scales to Understand the Critical Zone*, 2007, Elements, Vol. 3, pp. 307-314.

[Broadley et al., 2010] M.R. Broadley, J. Alcock, J. Alford, P. Cartwright, I. Foot, S.J. Fairweather-Tait, D.J. Hart, R. Hurst, P. Knott, S.P. McGrath, M.C. Meacham, K. Norman, H. Mowat, P. Scott, J.L. Stroud, M. Tovey, M. Tucker, P.J. White, S.D. Young, F.J.Zhao: *Selenium biofortification of high-yielding winter wheat (Triticum aestivum L.) by liquid or granular Se fertilisation*, 2010, Plant Soil, Vol. 332 pp. 5-18.

[Brookins, 1988] D.G. Brookins (Ed.): *Eh-pH Diagrams for Geochemistry*, 1988, Springer Berlin-Heidelberg.

[Brown & Shrift, 1982] T.A. Brown, A. Shrift: *Selenium: Toxicity And Tolerance In Higher Plants*, 1982, Biological Reviews, Vol. 57, pp. 65-81.

[Brozmanová et al., 2010] J. Brozmanová, D. Mániková, V. Vlčková, M. Chovanec: *Selenium: a double-edged sword for defense and offence in cancer*, 2010, Vol. 84, pp. 919-938.

[Buchner, Takahashi & Hawkesford, 2004] P. Buchner, H. Takahashi, M.J. Hawkesford: *Plant sulphate transporters: co-ordination of uptake, intracellular and long-distance transport*, 2004, Journal of Experimental Botany, Vol. 55, pp. 1765-1773.

[Burk & Hill, 1993] R.F. Burk, K.E. Hill: *Regulation of Selenoproteins*, 1993, Annual Review of Nutrition, Vol. 13, pp. 59-84.

[Bye, 1983] R. Bye: *Critical Examination of some common Reagents for Reducing Selenium Species in Chemical Analysis*, 1983, Talanta, Vol. 30, pp. 993-996.

[Bye & Lund, 1988] R. Bye, W. Lund: *Optimal conditions for the reduction of selenate to selenite by hydrochloric acid*, 1988, Fresenius' Journal of Analytical Chemistry, Vol. 332, pp. 242-224.

[Cantrell & Reeves, 2002] R.P. Cantrell, T.G. Reeves: *The Cereal of the World's Poor Takes Center Stage*, 2002, Science, Vol. 296 p. 53.

[Cappa et al., 2014] J.J. Cappa, C. Yetter, S. Fakra, P.J. Cappa, R. DeTar, C. Landes, E.A.H. Pilon-Smits, M.P. Simmons: *Evolution of selenium hyperaccumulation in Stanleya (Brassicaceae) as inferred from phylogeny, physiology and X-ray microprobe analysis*, 2014, New Phytologist, Vol. 285 pp. 583-595.

[Carey et al., 2012] A. Carey, K.G. Scheckel, E. Lombi, M. Newville, Y. Choi, G.J. Norton, A.H. Price, A.A. Meharg: *Grain Accumulation of Selenium*

Species in Rice (Oryza sativa L.), 2012, Environmental Science and Technology, Vol. 46, pp. 5557-5564.

[Catalano et al., 2008] J.G. Catalano, C. Park, P. Fenter, Z. Zhang: *Simultaneous inner- and outer-sphere arsenate adsorption on corundum and hematite*, 2008, Geochimica et Cosmochimica Acta, Vol. 72, pp. 1986 - 2004.

[Chan et al., 2009] Y.T. Chan, W.H. Kuan, T.Y. Chen, M.K. Wang: *Adsorption mechanisms of selenate and selenite on the binary oxide systems*, 2009, Water Research, Vol. 43 pp. 4412-4420.

[Chasteen & Bentley, 2003] T.G. Chasteen, R. Bentley: *Biomethylation of selenium and tellurium: microorganisms and plants*, 2003, Chemical Reviews, Vol. 103, pp. 1-26.

[Christensen et al., 1989] B.T. Christensen, F. Bertelsen, G. Gissel-Nielsen: *Selenite fixation by soil particle-size separates* , 1989, Journal of Soil Science, Vol. 40 p. 641-647.

[Clement et al., 1977] C.R. Clement, M.J. Hopper, L.H.P. Jones: *The Uptake of Nitrate by Lolium perenne from Flowing Nutrient Solution*, 1977, Journal of Experimental Botany, Vol. 29, pp. 453-464.

[Combs, 2001] G.F. Combs: *Selenium in global food systems*, 2001, British journal of Nutrition, Vol. 85, pp. 517-547.

[CO_2Now, 2014] CO2Now.org: *Earth's CO_2 Home Page*, 2014, http://co2now.org, (page visited 15:25, Jan. 19th, 2015).

[Dada et al., 2012] A.O. Dada, A.P. Olalekan, A.M. Olatunya, O. Dada: *Langmuir, Freundlich, Temkin and Dubinin-Radushkevich Isotherms Studies of Equilibrium Sorption of Zn^{2+} Unto Phosphoric Acid Modified Rice Husk*, 2012, IOSR Journal of Applied Chemistry, Vol. 3, pp. 38-45.

[Das et al., 2013] S. Das, M.J. Henry, J. Essilfie-Dughan: *Adsorption of selenate onto ferrihydrite, goethite, and lepidocrocite under neutral pH conditions*, 2013, Applied Geochemistry, Vol. 28, pp. 185-193.

[Davis & Kent, 1990] J.A. Davis, D.B. Kent: *Surface Complexation modelling in Aqueous Geochemistry*, 1990, in M.F. Hochella, Jr., A.F. White (Eds.): *Reviews in Mineralogy Vol. 23: Mineral-Water interface Geochemistry*, Mineralogical Society of America, Washington D.C., pp. 177-260.

[Dhillon & Dhillon, 2003] K.S. Dhillon, S.K. Dhillon: *Distribution and Management of Seleniferous Soils*, 2003, in D. Sparks (Ed.): *Advances in Agronomy*, 1st edn. Elsevier, Amsterdam, pp. 119-184.

[Dhillon et al., 2007] S.K. Dhillon, B.K. Hundal, K.S. Dhillon: *Bioavailability of selenium to forage crops in a sandy loam soil amended with Se-rich plant materials*, 2007, Chemosphere, Vol. 66, pp. 1734-1743.

[Dhillon & Dhillon, 2009] K.S. Dhillon, S.K. Dhillon: *Chemical Processes controlling the mobility of Se in soils*, 2009, in K.S. Dhillon, S.K. Dhillon: (Ed.): *Characterisation and management of Seleniferous Soils of Punjab*, Research Bulletin 1/2009, Ludhiana, pp. 13-15.

[Diener, 2012] A. Diener: *Structural Incorporation of Selenium into Pyrite and Mackinawite*, 2012, Dissertation, Karlsruhe Institute für Technologie, Karlsruhe.

[DIN EN ISO 12677:2013] *Chemische Analyse von feuerfesten Erzeugnissen durch Röntgenfluoreszenz-Analyse (RFA) – Schmelzaufschluss-Verfahren*, 2013.

[Dörfler, 2002] H. Dörfler: *Grenzflächen und kolloid-disperse Systeme: Physik und Chemie*, 2002, Springer, Berlin Heidelberg.

[Dove et al., 1997] M.T. Dove, D.A. Keen, A.C. Hannon, I.P. Swainson: *Direct measurement of the Si–O bond length and orientational disorder in the high-temperature phase of cristobalite*, 1997, Physical Chemistry of Minerals, Vol. 24, pp. 311-317.

[Duc et al., 2003] M. Duc, G. Lefevre, M. Federoff, J. Jeanjean, J.C. Rouchaud, F. Monteils-Rivera, J. Dumonceau, S. Milonjic: *Sorption of selenium anionic species on apatites and iron oxides from aqueous solutions*, 2003, Journal of environmental radioactivity, Vol. 70, pp. 61-72.

[Eiche, 2015] E. Eiche, F. Bardelli, A.K. Nothstein, L. Charlet, J. Göttlicher, R. Steininger, K.S. Dhillon, U.S. Sadana: *Selenium distribution and speciation in plant parts of wheat (Triticum aestivum) and Indian mustard (Brassica juncea) from a seleniferous area of Punjab, India*, 2015, Science of the Total Environment, Vol. 505, 952-961.

[Eurola et al., 2003] M. Eurola, G. Alfhan, A. Aro, P. Ekholm, V. Hietaniemi, H. Rainio, R. Rankanen, E.R. Venäläinen: *Results of Finnish Selenium Monitoring Program 2000-2001*, 2003, Agrifood Research Report 36; MTT Agrifood Research Finland, Jokioinen, 2003; p 42.

[FAOSTAT, 2012] FAOSTAT, Food and Agricultural Organisation of the United Nations (FAO), Statistics Division: *PART 3 – Feeding the world, Trends in the crop sector.*, 2012, pp. 182-197.

[Fernández-Martínez & Charlet, 2009] A. Fernández-Martínez, L. Charlet: *Selenium environmental cycling and bioavailability: a structural chemist point of view*, 2009, Reviews in Environmental Science and Biotechnology, Vol. 8, pp. 81-110.

[Ferrari & Renosto, 1972] G. Ferrari, F. Renosto: *Regulation of sulphate Uptake by Excises barley Roots in the Presence of Selenate*, 1972, Plant Physiology, Vol. 49, pp. 114-116.

[Finley, 2005] J.W. Finley: *Selenium Accumulation in Plant Foods*, 2005, Nutrition Reviews, Vol. 63, pp. 196-202.

[Flohé et al., 1973] L. Flohé, W.A. Günzler, H.H. Schock: *Glutathione peroxidase: a selenoenzyme*, 1973, FEBS letters, Vol. 32, pp. 132-134.

[Fordyce, 2007] F. Fordyce: *Selenium Geochemistry and Health*, 2007, AMBIO: a journal of the human environment, Vol. 36, pp. 94-97.

[Fordyce, 2013] F. Fordyce: *Selenium Deficiency and Toxicity in the Environment*, 2013, AMBIO: a journal of the human environment, Vol. 36, pp. 94-97.

[Fukushi & Sverjenski, 2007] K. Fukushi, D.A. Sverjenski: *A surface complexation model for sulphate and selenate on iron oxides consistent with spectroscopic and theoretical molecular evidence*, 2007, Geochimica et Cosmochimica Acta, Vol. 71, pp. 1-24.

[Geelhoed et al., 1997] J.S. Geelhoed, T. Hiemstra, W.H. van Riemsdijk: *Phosphate and sulphate adsorption on goethite: Single anion and competitive adsorption*, 1997, Geochimica et Cosmochimica Acta, Vol. 61, pp. 2389-2396.

[Ghose et al., 2010] S.K. Ghose, G.A. Waychunas, T.P. Trainor, P.J. Eng: *Hydrated goethite (α-FeOOH) (1 0 0) interface structure: Ordered water and surface functional groups*, 2010, Geochimica et Cosmochimica Acta, Vol. 74, pp. 1943-1953.

[Girling, 1984] C.A. Girling: *Selenium in agriculture and the environment*, 1984, Agriculture, Ecosystems and Environment, Vol. 11, pp. 37-65.

[Gissel-Nielsen, 1973] G. Gissel-Nielsen: *Uptake and Distribution of Added Selenite and Selenate by Barley and Red Clover as Influenced by Sulphur*, 1973, Journal of the Science of Food and Agriculture, Vol. 24, pp. 649-655.

[Glass, 1954] H.D. Glass: *High-Temperature Phases from Kaolinite and Halloysite*, 1954, American Mineralogist, Vol. 39, pp. 193-207.

[Goff et al., 2002] S.A. Goff, D. Ricke, T. Lan, G. Presting, R. Wang, M. Dunn, J. Glazebrook, A. Sessions, P. Oeller, H. Varma, (. . .) S. Briggs: *A Draft Sequence of the Rice Genome (Oryza sativa L. ssp. japonica)*, 2002, Science, Vol. 296, pp. 92-100.

[Goldberg et al., 2007] S. Goldberg, L.J. Criscentib, D.R. Turner, J. A. Davis, K.J. Cantrelle: *Adsorption-Desorption Processes in Subsurface Reactive Transport Modelling*, 2007, Vadose Zone Journal, Vol. 6, pp. 407-435.

[Goldberg et al., 2013] S. Goldberg: *modelling Selenite Adsorption Envelopes on Oxides, Clay Minerals, and Soils using the Triple Layer Model*, 2013, Soil Science Society of America Journal, Vol. 77, pp. 64-71.

[Gustafsson & Johnsson, 1992] J.P. Gustafsson, L. Johnsson: *Selenium retention in the organic matter of Swedish forest soil*, 1992, Journal of Soil Science, Vol. 43, pp. 461-472.

[Hagarová, Žemberyová & Bajčan, 2005] I. Hagarová, M. Žemberyová, D. Bajčan: *Sequential and Single Step Extraction Procedures Used for Fractination of Selenium in Soil Samples*, 2005, Chemical Papers, Vol. 59, pp. 93-98.

[Hartikainen, 2005] H. Hartikainen: *Biogeochemistry of selenium and its impact on food chain quality and human health*, 2005, Journal of Trace Elements in Medicine and Biology, Vol. 18, pp. 309-318.

[Haug et al., 2007] A. Haug, R.D. Graham, O.A. Christopherson, G.H. Lyons: *How to use the world's scarce selenium resources efficiently to increase the selenium concentration in food*, 2007, Microbial Ecology in Health and Disease, Vol 19, pp. 209-228.

[Hayes et al., 1987] K.F. Hayes, A.L. Roe, G.E. Brown, Jr., K.O. Hodgson, J.O. Leckie, G.A. Parks: *In-situ X-ray absorption study of surface complexes: Selenium oxy-anions on FeOOH*, 1987, Science, Vol. 238, pp. 783-786.

[Haygarth, 1994] P.M. Haygarth: *Global Importance and Global Cycling of Selenium*, 1994, in W.T. Frankenberger, Jr., S. Benson (Eds.): *Selenium In The Environment*, Marcel Dekker, Inc., Hong Kong, pp. 1-28.

[Hiemstra & van Riemsdijk, 1995] T. Hiemstra, W.H. van Riemsdijk: *A Surface Structural Approach to Ion Adsorption: The Charge Distribution (CD) Model*, 1995, Journal of Colloid and Interface Science, Vol. 179, pp. 488-508.

[Hingston et al. 1971] F.J. Hingston, A.M. Posner, J.P. Quirk: *Competitive Adsorption of Negatively Charged Ligands on Oxide Surfaces*, 1971, Discussions of the Faraday Society, Vol. 52, pp. 334-342.

[Hochella & White, 1990] M.F. Hochella, Jr., A.F. White: *Mineral-Water interface Geochemistry: An Overview*, 1990, in M.F. Hochella, Jr., A.F. White (Eds.): *Reviews in Mineralogy Vol. 23: Mineral-Water interface Geochemistry*, Mineralogical Society of America, Washington D.C., pp. 1-5.

[Holben & Smith, 1999] D.H. Holben, A.M. Smith: *The diverse role of selenium within selenoproteins: a review*, 1999, Journal of the American Dietetic Association, Vol. 99, pp. 836-843.

[Hopper & Parker, 1999] J.L. Hopper, D.R. Parker: *Plant availability of selenite and selenate as influenced by the competing ions phosphate and sulphate*, 1999, Plant and Soil, Vol. 210, pp. 199-207.

[Horneck et al., 2011] D.A. Horneck, D.M. Sullivan, J.S. Owen, J.M. Hart: *Soil Test Interpretation Guide*, 2011, Oregon State University Extension Service, EC1478.

[Howard, 1977] J.H. Howard: *Geochemistry of selenium: formation of ferroselite and selenium behavior in the vicinity of oxidizing sulfide and uranium deposits*, 1977, Geochimica et Cosmochimica Acta, Vol. 41, pp. 1665-1678.

[Hu et al., 2002] Q. Hu, L. Chen, J. Xu, Y. Zhang, G. Pang: *Determination of selenium concentration in rice and the effect of foliar application of Se-enriched fertiliser or sodium selenite on the selenium content of rice*, 2002, Journal of the Science of Food and Agriculture, Vol. 82, pp. 869-872.

[Huang et al., 2013] Y. Huang, Q. Wang, J. Gao, Z. Lin, G.S. Bañuelos, L. Yuan, Z. Yin: *Daily Dietary Selenium Intake in a High Selenium Area of Enshi, China*, 2013, Nutrients, Vol. 5, pp. 700-710.

[Hur & Schlautmann, 2003] J. Hur, M. Schlautmann: *Molecular weight fractionation of humic substances by adsorption onto minerals*, 2003, Journal of Colloid and Interface Science, Vol. 64, pp. 313-321.

[Imai, 1998] H. Imai, K. Narashima, M. Arai, H.Sakamoto, N. Chiba, Y. Nakagawa: *Suppression of Leukotriene Formation in RBL-2H3 Cells That Overexpressed Phospholipid Hydroperoxide Glutathione Peroxidase*, 1998, The Journal of Biological Chemistry, Vol. 273, pp.1990-1997.

[Institute of Medicine, 2000] Institute of Medicine: *Selenium*, 2000, in Panel on Dietary Antioxidants and Related Compounds (Ed.): *Dietary reference intakes for vitamin C, vitamin E, selenium, and carotenoids*, National Academy Press, Washington, pp. 284-324.

[Izawa & Shimamoto, 1996] T. Izawa & K. Shimamoto: *Becoming a model plant: The importance of rice to plant science* , 1996, Trends in Plant Science, Vol. 1, pp. 95-99.

[Jiyun et al., 1982] L. Jiyun, R. Shangxue, C. Daizhong: *A Study of Kashin-Beck Disease Associated with Environmental Selenium in Shanxi Area*, 1982, Acta Scientiae Circumstantiae, Vol. 2, pp. 91-101.

[Kálmán, 1971] A. Kálmán: *Mean X–O bond lengths of the XO 4 n–tetrahedral oxy-anions*, 1971, Journal of the Chemical Society A: Inorganic, Physical, Theoretical, Vol. 1, pp. 1857-1859.

[Kang et al., 1991] Y. Kang, H. Yamada, K. Kyuma, T. Hattori, S. Kigasawa: *Selenium in Soil Humic acid*, 1991, Soil Science Plant Nutrition, Vol. 37, pp. 2441-248.

[Keskinen et al., 2010] R. Keskinen, M. Turakainen, H. Hartikainen: *Plant availability of soil selenate additions and selenium distribution within wheat and ryegrass*, 2010, Plant and Soil, Vol. 333, pp. 301-313.

[Khan & Hell, 2014] M.S. Khan, R. Hell: *Applied Cell Biology of Sulphur and Selenium in Plants*, in P. Nick, Z. Opatrny (eds.) *Applied Plant Cell Biology*, 2014, Springer, Berlin Heidelberg, pp. 247-272.

[Kikkert et al., 2013] J. Kikkert, B. Hale, E. Berkelaar: *Selenium accumulation in durum wheat and spring canola as a function of amending soils with selenite, selenate and or sulphate*, 2013, Plant Soil , Vol. 372, pp. 629-641.

[Kinjo & Pratt, 1971] T. Kinjo, P.F. Pratt: *Nitrate Adsorption: II. In Competition with Chloride, sulphate, and Phosphate*, 1971, Soil Science Society of America Journal, Vol. 35, pp. 725-728.

[Kosmulski, 2002] M. Kosmulski: *The pH-Dependent Surface Charging and the Points of Zero Charge*, 2002, Journal of Colloid and Interface Science, Vol. 253, pp. 77-87.

[Kulp & Pratt, 2004] T. Kulp, L. Pratt: *Speciation and weathering of selenium in upper cretaceous chalk and shale from South Dakota and Wyoming, USA*, 2004, Geochimica et Cosmochimica Acta, Vol. 68, pp. 3687-3701.

[Lakin, 1972] H. W. Lakin: *Selenium accumulation in soils and its absorption by plants and animals*, 1972, Geological Society of America Bulletin, Vol. 83, pp. 181-190.

[Lagaly & Köster, 1993] G. Lagaly, H.M. Köster: *Tone und Tonminerale*, 1993, in G. Lagaly, H.M. Köster (Eds.): *Tonminerale und Tone - Struktur, Eigenschaften, Anwendungen und Einsatz in Industrie und Umwelt*, Steinkopff Verlag, Darmstadt, pp. 1-32.

[Langmuir, 1918] I. Langmuir: *The adsorption of gases on plane surfaces of glass, mica and platinum*, 1918, Journal of the American Chemical Society, Vol. 40, pp. 1361-1403.

[Läuchli, 1993] A. Läuchli: *Selenium in Plants: Uptake, Functions, and Environmental Toxicity*, 1993, Botanica Acta, Vol. 106, pp. 455-485.

[LeDuc et al., 2004] D.L. LeDuc, A.S. Tarun, M. Montes-Bayon, J. Meija, M.F. Malit, C.P. Wu, M. AdelSamie, C. Chiang, A. Tagmount, M. deSouza, B. Neuhierl, A. Böck, J. Caruso, N. Terry: *Overexpression of selenocysteine methyltransferase in Arabidopsis and Indian mustard increases selenium tolerance and accumulation*, 2004, Plant Physiology, Vol. 135, pp. 377-383.

[Leggett & Epstein, 1956] J.E. Leggett, E. Epstein: *Kinetics of sulphate absorption by barley roots*, 1956, Plant Physiology , Vol. 31, pp. 222-226.

[Lemly, 1993] A.D. Lemly: *Guidelines for evaluating selenium data from aquatic monitoring and assessment studies*, 1993, Environmental Monitoring and Assessment, Vol. 28, pp. 83-100.

[Lenz & Lens, 2009] M. Lenz, P.N.L. Lens: *The essential toxin: The changing perception of selenium in environmental sciences*, 2009, Science of the Total Environment, Vol. 407, pp. 3620-3633.

[Levander & Burk, 2006] O.A. Levander, R.F. Burk: *Update of human dietary standards for selenium*, 2006, in D.L. Hatfield, M.J. Berry, V.N. Gladyshev

(Eds.): *Selenium: Its molecular biology and role in human health*, 2nd edn., Springer, New York, pp 399-410.

[Lévesque, 1974] M. Lévesque: *Selenium distribution in Canadian soil profiles*, 1974, Canadian Journal of Soil Science, Vol. 54, pp. 63-68.

[Li et al., 2007] H. Li, S.P. McGrath, F. Zhao: *Selenium uptake, translocation and speciation in wheat supplied with selenate or selenite*, 2007, New Phytologist, Vol. 178, pp. 92-102.

[Li et al., 2010] H. Li, E. Lombi, J.L. Stroud, S.P. McGrath, F. Zhao: *Selenium speciation in soil and rice - influence of water management and selenium fertilisation*, 2010, Journal of Agricultural Food Chemistry, Vol. 58, pp. 11837-11843.

[Liu & Gu, 2009] K. Liu, Z. Gu: *Selenium Accumulation in Different Brown Rice Cultivars and Its Distribution in Fractions*, 2009, Journal of Agricultural and Food Chemistry , Vol. 57, pp. 695-700.

[Lo & Chen, 1997] S. Lo, T. Chen: *Adsorption of Se(IV) and Se(VI) on an iron-coated sand from water*, 1997, Chemosphere, Vol. 35, pp. 919-930.

[Lombeck et al., 1987] I. Lombeck, H. Menzel, D. Frosch: *Acute selenium poisoning of a 2-year-old child*, 1987, European Journal of Pediatrics, Vol. 146, pp. 308-312.

[Lussier et al., 2003] C. Lussier, V. Veiga, S. Baldwin: *The geochemistry of selenium associated with coal waste in the Elk River Valley, Canada*, 2003, Environmental Geology, Vol. 35, pp. 919-930.

[Lützelschwab, 1990] M. Lützelschwab: *Biochemische und immunologische Charakterisierung von Funktionen der Plasmamembran in Curcubita pepo L.: Evidenz für Heterogenität der Plasmamembran*, 1990, Dissertation, Albert-Ludwig-Universität, Freiburg.

[Manceau & Charlet, 1994] A. Manceau, L. Charlet: *The mechanism of selenate adsorption on goethite and hydrous ferric oxide*, 1994, Journal of Colloid and Interface Science, Vol. 168, pp. 87-93.

[Marx et al., 1999] E.S. Marx, J. Hart, and R.G. Stevens: *Soil Test Interpretation Guide*, 1999, Oregon State University Extension Service, EC1478.

[Minaev et al., 2005] V.S.Minaev, S.P. Timoshenkov, V.V. Kalugin: *Structural and phase transformations in condensed selenium*, 2005, Journal of Optoelectronics and Advanced Materials, Vol. 7, pp. 1717-1741.

[Miranda-Trevino & Coles, 2003] J.C. Miranda-Trevino, C.A. Coles: *Kaolinite properties, structure and influence of metal retention on pH*, 2003, Applied Clay Science, Vol. 23, pp. 133-139.

[Misra, 2010] S. Misra, D. Peak, S. Niyogi: *Application of XANES spectroscopy in understanding the metabolism of selenium in isolated rainbow trout hepatocytes: insights into selenium toxicity*, 2010, Metallomics, Vol. 2, pp. 710-717.

[Moreno-Reyes et al., 1998] R. Moreno-Reyes, C. Suetens, F. Mathieu, F. Begaux, D. Zhu, M. T. Rivera, M. Boelaert, J. Nève, N. Perlmutter, J. Vanderpas: *Kashin-Beck Osteoarthropathy in Rural Tibet in Relation to Selenium and Iodine Status*, 1998, The New England Journal of Medicine, Vol. 339, pp. 1112-1120

[Mortimer, 2010] C. Mortimer, U. Müller: *Die Elemente der 5 Hauptgruppe*, 2010, in C. Mortimer, U. Müller (Eds.): *Chemie - Basiswissen der Chemie*, 10th edn., Georg Thieme Verlag, Stuttgart, pp. 421-445.

[Murashige & Skoog, 1962] T. Murashige, F. Skoog: *A Revised Medium For Rapid Growth And Bio Assays With Tabbacco Tissue Cultures*, 1962, Physiologia Plantarum, Vol. 15, pp. 473-497.

[Mikkelsen & Wan, 1990] R.L. Mikkelsen, H.F. Wan: *The effect of selenium on sulfur uptake by barley and rice*, 1990, Plant and Soil, Vol. 121, pp. 151-153.

[Navarro-Alarcon & Cabrera-Vique, 2008] M. Navarro-Alarcon, C. Cabrera-Vique: *Selenium in food and the human body: A review*, 2008, Science of the Total Environement, Vol. 400, pp. 115-141.

[Neal, 1995] R.H. Neal: *Chapter 12. Selenium*, 1995, in B.J. Alloway (Ed.): *Heavy Metals In Soils*, Blackie Academic and Professional, London, pp. 260-283.

[Nesterov, 1964] J. Trofast: *Berzelius' Discovery of Selenium*, 2011, Chemistry International, Vol. 33, pp. 16-19.

[Newville, 2004] M. Newville: *Fundamentals of XAFS*, 2004, http://xafs.org/Tutorials?action=AttachFile&do=view&target=Newville_xas_fundamentals.pdf, (page visited 15:18, Mar. 15th, 2015).

[O'Toole & Raisbeck, 1964] D. O'Toole, M.F. Raisbeck: *Pathology of experimentally induced chronic selenosis (alkali disease) in yearling cattle*, 1995, Journal of Veterinary Diagnostic, Vol. 7, pp. 364-373.

[Oger et al., 2004] P.M. Oger, I. Daniel, B. Cournoyer, A. Simionovici *In situ micro X-ray absorption near edge structure study of microbiologically reduced selenite (SeO_2^{2-})* , 2004, Spectrochimica Acta Part B, Vol. 59, pp. 1681-1686.

[Ogo et al., 2007] Y. Ogo, R.N. Itai, H. Nakanishi, T. Kobayashi, M. Takahashi, S. Mori, N.K. Nishizawa: *The rice bHLH protein OsIRO2 is an essential regulator of the genes involved in Fe uptake under Fe-deficient conditions*, 2007, The Plant Journal, Vol. 51, pp. 366-377.

[Ohlendorf, 2002] H.M. Ohlendorf: *The birds of Kesterson Reservoir: a historical perspective*, 2002, Aquatic Toxicology, Vol. 57, pp. 1-10.

[Oldfield, 2006] J.E. Oldfield: *Chapter 1. Selenium: A historical perspective*, 2006, in D.L. Hatfield, M.J. Berry, V.N. Gladyshev (Eds.): *Selenium: Its molecular biology and role in human health*, 2nd edn. Springer, New York, pp. 1-6.

[Oldfield, 1987] J.E. Oldfield: *The Two Faces of Selenium*, 1987, Journal of Nutrition, Vol. 117, pp. 2002-2008.

[Oyebanji et al., 2009] O. B. Oyebanji, O. Nweke, O. Odebunmi, N. B. Galadima, M. S. Idris, U. N. Nnodi, A. S. Afolabi, G. H. Ogbadu: *Simple, effective and economical explant-surface sterilisation protocol for cowpea, rice and sorghum seeds*, 2009, African Journal of Biotechnology, Vol. 8, pp. 5395-5399.

[Oliver, 1997] M. A. Oliver: *Soil and human health: a review*, 1997, European Journal of Soil Science, Vol. 48, pp. 573-592.

[Papp, 2007] L.V. Papp, J. Lu, A. Holmgren, K.K. Khanna: *From Selenium to Selenoproteins: Synthesis, Identity, and Their Role in Human Health*, 2007, Antioxidants & Redox Signaling, Vol. 9, pp. 775-806.

[Parks, 1990] G.A. Parks: *Surface Energy and Adsorption at Mineral-Water Interfaces: An Introduction*, 1990, in M.F. Hochella, Jr., A.F. White (Eds.): *Reviews in Mineralogy Vol. 23: Mineral-Water interface Geochemistry*, Mineralogical Society of America, Washington D.C., pp. 133-175.

[Paul & Pryor, 1972] G.L. Paul, A.W. Pryor: *The Study of Sodium Nitrate by Neutron Diffraction*, 1972, Acta Crystallographica Section B: Structural Crystallography and Crystal Chemistry, Vol. 28, pp. 2700-2702.

[Peak, 2006] D. Peak: *Adsorption mechanisms of selenium oxy-anions at the aluminum oxide/water interface*, 2006, Journal of Colloid and Interface Science, Vol. 303, pp. 337-345.

[Pickering et al., 1999] I.J. Pickering, G.N. George, V. Van Fleet-Stalder, T.G. Chasteen, R.C. Prince: *X-ray absorption spectroscopy of selenium-containing amino acids*, 1999, Journal of Biological Inorganic Chemistry, Vol. 4, pp. 791-794.

[Pilon-Smits et al., 2009] E. Pilon-Smits E., C.F. Quinn, W. Tapken, M. Malagoli, M. Schiavon: *Physiological functions of beneficial elements.*, 2009, Current Opinion in Plant Biology, Vol. 12, pp. 267-274.

[Presser, 1994] T.S. Presser: *The Kesterson Effect*, 1994, Environmental Management, Vol. 18, pp. 437-454.

[Presser & Luomo, 2009] T.S. Presser, S.N. Luoma: *modelling of Selenium for the San Diego Creek Watershed and Newport Bay, California*, 2009, U.S. Geological Survey Open-File Report 2009-1114, Reston, Virginia.

[Presser & Ohlendorf, 1987] T.S. Presser, H.M. Ohlendorf: *Biogeochemical Cycling of Selenium in the San Joaquin Valley, California, USA*, 1987, Environmental Management, Vol. 11, pp. 805-821.

[Presser & Piper, 1998] T.S. Presser, D.Z. Piper: *Mass balance approach to selenium cycling through the San Joaquin Valley: From source to river to bay* 1998, in W.T. Frankenberg, R.A. Ingberg (Eds.) *Environmental chemistry of selenium*, 1998, Marcel Dekker, Inc. New York, pp.153-182.

[Quinn et al., 2008] C.F. Quinn, J.L. Freeman, M.L. Galeas, E.M. Klamper, E.A.H. Pilon-Smitts: *The role of selenium in protecting plants against prairie dog herbivory: implications for the evolution of selenium hyperaccumulation*, 2008, Oecologia, Vol. 155, pp. 267-275.

[Rajan & Watkinson, 1976] S.S.S. Rajan, J.H. Watkinson: *Adsorption of Selenite and Phosphate on an Allophane Clay*, 1976, Soil Science Society of America Journal, Vol. 40, pp. 51-54.

[Ramesh & Riyazuddin, 2010] K. A. Ramesh, P. Riyazuddin: *Chemical Interferences in hydride generation atomic spectrometry*, 2010, Trends in Analytical Chemistry, Vol. 29, p. 166-176.

[Rayman, 2000] M.P. Rayman: *The importance of selenium to human health*, 2000, The Lancet, Vol. 356, pp. 233-241.

[Rayman, 2008] M.P. Rayman: *Food-chain selenium and human health: emphasis on intake*, 2008, British Journal of Nutrition, Vol. 100, pp. 254-268.

[Reamer & Zoller, 1980] D.C. Reamer, W.H. Zoller: *Selenium Biomethylation Products from Soil and Sewage Sludge*, 1980, Science, Vol. 208, pp. 500-502.

[Reeder et al., 2006] R.J. Reeder, M.A.A. Schoonen, A. Lanzirotti: *Metal speciation and its role in bioaccessibility and bioavailability*, 2006, in J.J. Rosso (Ed.): *The emergent field of medical mineralogy and geochemistry*, 1st edn. Mineralogical Society of America and the Geochemical Society, New York, pp 59–113.

[Reilly, 1997] C. Reilly: *Selenium Metabolism* 1997, in C. Reilly (Ed.) *Selenium in Food and Health*, 1997, Blackie Academic & Professional, London, pp.43-66.

[Rietra et al., 2001] R.P.J.J. Rietra, T. Hiemstra, W.H. van Riemsdijk: *Comparison of Selenate and sulphate Adsorption on Goethite*, 2001, Journal of Colloid and Interface Science, Vol. 260, pp. 384-390.

[Rosenfeld & Beath, 1964] I. Rosenfeld, O.A. Beath: *Selenium. Geobotany, biochemistry, toxicity, and nutrition.* 1964, Academic Press, New York.

[Rotruck et al., 1973] J. T. Rotruck, A.L. Pope, H.E. Ganther, A.B. Swanson, D.G. Hafeman, W.G.Hoekstra: *Selenium: Biochemical Role as a Component of Glutathione Peroxidase*, 1973, Science, Vol. 179, pp. 588-590.

[Rovira et al., 2008] M. Rovira, J. Giménezm María Martínez, Xavier Martínez-Lladó, J. de Pablo, V. Martí, L. Duro: *Sorption of selenium(IV) and selenium(VI) onto natural iron oxides: Goethite and hematite* 2008, Journal of Hazardous Materials, Vol. 150, pp. 279-284.

[Ryser et al., 2005] A.L.Ryser, D.G. Strawn, M.A. Marcus, J.L. Johnson-Maynard, M.E. Gunter, G. Möller: *Micro-spectroscopic investigation of selenium-bearing minerals from the Western US Phosphate Resource Area*, 2005, Geochemical Transactions, Vol. 1, pp. 1-11.

[Schroth & Sposito, 1997] B.K. Schroth, G. Sposito: *Surface charge properties of kaolinite*, 1997, Clays and Clay Minerals, Vol. 45, pp. 85-91.

[Schwartz & Foltz, 1957] K. Schwartz, C.M. Foltz: *Selenium as an integral part of factor 3 against dietary necrotic liver degeneration*, 1957, Journal of the American Chemical Society, Vol. 79, pp. 3292-3293.

[Séby et al., 2001] F. Séby, M. Potin-Gautier, E. Giffaut, G. Borge, O.F.X. Donard: *A critical review of thermodynamic data for selenium species at 25°C*, 2001, Chemical Geology, Vol. 171, pp. 173-194.

[Pezzarossa & Petruzelli, 2001] B. Pezzarossa, G. Petruzelli, H.M. Selim, D.L. Sparks: *Selenium contamination in soil: sorption and desorption processes*, 2001, in H.M. Selim, D.L. Sparks (Eds.): *Heavy metals release in soils*, CRC Press, New York, pp. 191-206.

[Singh & Gilkes, 1998] B. Singh, R.J. Gilkes: *Properties of soil kaolinites from south-western Australia*, 1992, Journal of Soil Science, Vol. 43, pp. 645-667.

[Singh et al., 1981] M. Singh, N. Singh, P.S. Relan: *Adsorption and Desorption of Selenite and Selenate Selenium on Different Soils*, 1981, Soil Science, Vol. 132, pp. 134-141.

[Smith & Watkinson, 1984] G.S. Smith, J.H. Watkinson: *Selenium toxicity in perennial ryegrass and white clover*, 1984, New Phytologist, Vol. 97, pp. 557-564.

[Sors et al., 2005] T.G. Sors, D.R. Ellis, D.E. Salt: *Selenium uptake, translocation, assimilation and metabolic fate in plants*, 2005, Photosynthesis Research, Vol. 86, pp. 373-389.

[Spallholz & Hoffman, 2002] J.E. Spallholz, D.J. Hoffman: *Selenium toxicity: cause and effects in aquatic birds*, 2002, Aquatic Toxicology, Vol. 57, pp. 27-37.

[Sposito, 1984] G. Sposito: *The surface Chemistry of Soils*, 1984, in G. Sposito (Ed.): *The Surface Chemistry of Soils*, Oxford University Press, New York, pp. 106-108.

[Sposito, 1998] G. Sposito: *On points of zero charge*, 1998, Environmental Science and Technology, Vol. 32, pp. 2815-2819.

[Sposito, 2004] G. Sposito: *The surface chemistry of natural particles*, 2004, Oxford University Press, New York.

[Stelling, 2010] M. Stelling: *Stable iron isotope variations in higher plants - systematics and controls*, 2010, Dissertation, Gottfried Wilhelm Leibniz Universität Hannover.

[Stolz & Oremland] J.F. Stolz, R.S. Oremland: *Bacterial respiration of arsenic and selenium*, 1999, Federation of European Microbiological Societies Micorbiology Reviews, Vol. 23, pp. 615-627.

[Stumm & Morgan, 1996] W. Stumm, J. Morgan: *Aquatic Chemistry - Chemical Rate and Equilibria in natural waters*, 1996, Wiley Inter-Science, New York.

[Stwertka, 2002] A. Stwertka: *Selenium*, 2002, in A. Stwertka (Ed.): *A Guide To The Elements* 2nd edn., Oxford University Press, New York, pp. 106-108.

[Su & Suarez, 2000] C. Su, D.L. Suarez: *Selenate and Selenite Sorption on Iron Oxides: An Infrared and Electrophoretic Study*, 2000, Soil Science, society America Journal, Vol. 64, pp. 101-111.

[Sun et al., 2010] G. Sun, X. Liu, P.N. Williams, Y. Zhu: *Distribution and Translocation of Selenium from Soil to Grain and its Speciation in Paddy Rice (Oryza sativa L.*, 2010, Environmental Science and Technology, Vol. 44, pp. 6706-6711.

[Tamai & Ma, 2003] K. Tamai, J.F. Ma: *Characterisation of silicon uptake by rice roots*, 2003, New Phytologist, Vol. 158, pp. 431-436.

[Terry et al., 2000] N. Terry, A.M. Zayed, M.P. de Souza, A.S. Tarun: *Selenium in Higher Plants*, 2000, Annual Review of Plant Physiology and Plant Molecular Biology, Vol. 51, pp. 401-432.

[Tombácz & Szekeres, 2006] E. Tombácz, M. Szekeres: *Surface charge heterogeneity of kaolinite in aqueous suspension in comparison with montmorillonite*, 2006, Applied Clay Science, Vol. 34, pp. 105-124.

[Tokunaga et al., 1994] T.K. Tokunaga, S.R. Sutton, S. Bajt: *Mapping of Selenium Concentrations in Soil Aggregates with Synchrotron X-ray Fluorescence Microprobe*, 1994, Soil Science, Vol. 158, pp. 421-434.

[Tributh, 1976] H.. Tributh: *Die Umwandlung glimmerartiger Schichtsilikate zu aufweitbaren Dreischichttonmineralen*, 1976, Zeitschrift für Pflanzenernährung und Bodenkunde, Vol. 139, pp. 7-25.

[Trofast, 2011] J. Trofast: *Berzelius' Discovery of Selenium*, 2011, Chemistry International, Vol. 33, pp. 16-19.

[Tzeng & Zeitlin, 1978] J. Tzeng, H. Zeitlin: *The separation of selenium from sea water by Adsorption colloid flotation*, 1978, Analytica Chimica Acta, Vol. 101, pp. 71-77.

[Ulrich & Shrift, 1968] J.M. Ulrich, A. Shrift: *Selenium Absorption by Execised Astralagus Roots*, 1968, Plant Physiology, Vol. 43, pp. 14-20.

[Virupaksha & Shrift, 1965] T.K. Virupaksha, A. Shrift: *Biogeochemical differences between selenium accumulator and non-accumulator Astragalus species* 1965, Biochimica Biophysica Acta, Vol. 107, pp. 69-80.

[Wang & Gao, 2001] Z.J. Wang, Y.X. Gao: *Biogeochemical cycling of selenium in Chinese environments*, 2001, Applied Geochemistry, Vol. 16, pp. 1345-1351.

[Wang et al. 2013] P. Wang, N.W. Menzies, E. Lombi, B.A. McKenna, M.D. de Jonge, D.J. Paterson: *In situ speciation and distribution of toxic selenium in hydrated roots of cowpea*, 2013, Plant Physiology, Vol. 163, pp. 407-418.

[Water Education Foundation, 2014] Water Education Foundation: *Kesterson Reservoir*, 2014, http://www.watereducation.org/aquapedia/kesterson-reservoir, (page visited 20:25, Jan. 18th, 2015).

[Wen & Carignan, 2007] H. Wen, J. Carignan: *Reviews on atmospheric selenium: Emissions, speciation and fate*, 2007, Atmospheric Environment, Vol. 41, pp. 7151-7165.

[Wijnja & Schulthess, 2000] H. Wijnja, C.P. Schulthess: *Vibrational Spectroscopy Study of Selenate and sulphate Adsorption Mechanisms on Fe an Al (Hydr)oxide Surfaces*, 2000, Journal of Colloid and Interface Science, Vol. 229, pp. 286-297.

[Wijnja & Schulthess, 2002] H. Wijnja, C.P. Schulthess: *Effect of Carbonate on the Adsorption of Selenate and sulphate on Goethite*, 2002, Soil Science American Journal, Vol. 66, pp. 1190-1197.

[Winkel et al., 2011] L.H.E. Winkel, C.A. Johnson, M. Lenz, T. Grundl, O.X. Leupin, M. Amini, L. Charlet: *Environmental Selenium Research: From Microscopic Processes to Global Understanding*, 2011, Environmental Science And Technology, Vol. 46, pp. 571-579.

[Wu et al., 2000] C. Wu, S. Lo, C. Lin: *Competitive adsorption of molybdate, chromate, sulphate, selenate, and selenite on* γ*-*Al_2O^3, 2000, Colloids and Surfaces A: Physicochemical and Engineering Aspects, Vol. 166, pp. 251-259.

[Xue et al., 2001] T. Xue, H. Hartikainen, V. Piironen: *Antioxidative and growth-promoting effects of selenium on senescing lettuce*, 2001, Plant and Soil, Vol. 61, pp. 237-255.

[Xu & Hu, 2004] J. Xu, Q. Hu: *Effect of foliar application of Selenium on antioxidant activity of aqueous and ethanolic extracts of Selenium-enhanced rice*, 2004, Journal of Agricultural and Food Chemistry, Vol. 52, pp. 1759-1763.

[Yang et al., 2011] S.I. Yang, J.R. Lawrence, G.D.W. Swerhone, I.J. Pickering: *Biotransformation of selenium and arsenic in multi-species biofilm*, 2011, Environmental Chemistry, Vol. 8, pp. 543-551.

[Young et al., 2009] T.F. Young, K. Finley, W.J. Adams, J. Besser, W.D. Hopkins, D. Jolley, E. McNaughton, T.S. Presser, D.P. Shaw, J. Unrine: *What you need to know about selenium*, 2009, in P.M. Chapman, W.J. Adams, M.L. Brooks, C.D. Gelos, S.N. Luoma, W.A. Maher, H.M. Ohlendorf, T.S. Presser, D.P. Shaw (Eds.) *Ecological Assessment of Selenium in the Aquatic Environment*, CRC Press, New York, pp. 7-45.

[Yukselen & Kaya, 2002] Y. Yukselen, A. Kaya: *Zeta Potential of Kaolinite in the Presence of Alkali, Alkaline Earth and Hydrolyzable Metal Ions*, 2002, Water, Air and Soil Pollution, Vol. 145, pp. 155-168.

[Zayed et al., 1998] A. Zayed, C.M. Lytle, N. Terry: *Accumulation and volatilisation of different chemical species of selenium by plants*, 1998, Planta, Vol. 206, pp. 284-292.

[Zhang & Sparks, 1990] P. Zhang, D.L. Sparks: *Kinetics of Selenate and SeleniteAdsorption/Desorption at the Goethite/Water Interface*, 1990, Environmental Science and Technology, Vol. 24, pp. 1848-1856.

[Zhang & Moore, 1996] Y.Q. Zhang, J.N. Moore: *Selenium fractionation and speciation in a wetland system*, 1996, Environmental Science and Technology, Vol. 30, pp. 2613-2619.

[Zhang et al., 2003] J. Zhang, G. Pan, J. Chen, Q. Hu: *Uptake and transport of selenite and selenate by soybean seedlings of two genotypes*, 2003, Plant and Soil, Vol. 253, pp. 437-443.

[Zhang et al., 2005] E.S. Zhang, H.L. Wang, X.X. Yan, L.D. Zhang: *Comparison of short-term toxicity between Nano-Se and selenite in mice*, 2005, Life Science, Vol. 76, pp. 1099-1109.

[Zhang et al., 2006] L.H. Zhang, W. Shi, X. Wang: *Difference in selenite absorption between high- and low-selenium rice cultivars and its mechanism*, 2006, Plant and Soil, Vol. 282, pp. 183-193.

[Zhao et al., 2010] X.Q. Zhao, N. Mitani, N. Yamaji, R.F. Shen, J.F. Ma: *Involvement of Silicon Influx Transporter OsNIP2;1 in Selenite Uptake in Rice*, 2010, Plant Physiology, Vol. 153, pp. 1871-1877.

[Zhu et al., 2008] Y. Zhu, E.A.H. Pilon-Smitts, F. Zhao, P. WIlliams, A.A. Meharg: *Selenium in higher plants - understanding mechanisms for biofortification and phytoremediation*, 2008, Trends in Plant Science, Vol. 14, pp. 436-442.

[Zieve & Peterson, 1984] R. Zieve, P.J. Peterson: *The accumulation and assimilation of dimethylselenide by four plant species*, 1984, Planta, Vol. 160, pp. 180-184.

Appendix

Due to difficulties in downsizing the document from DIN A4 to DIN A5, the appendix was not included in this book as it contains multiple tables spanning DIN A4 pages in tiny fontsizes, which would be illegible in DIN A5.

The original appendix contains:

- Phreeq-C Input files
- the complete photo documentation of all harvested plants
- plant Se content measurements: HG-FIAS (analytical raw data)
- nutrient solution composition measurements: IC, ICP-MS (analytical raw data)
- material characterisation data for kaolinite & goethite: SEM, XRD, TG-DSC
- sorption solution composition measurements: IC, ICP-MS (analytical raw data)
- ANKA data: all XANES spectra measured at FLUO and SUL-X with respective microscope images
- detailed linear combination fitting results

The electronic publication of this dissertation complete with color images and the full appendix can be found at:

http://www.bibliothek.kit.edu
DOI(KIT): 10.5445/IR/1000049456
URN: urn:nbn:de:swb:90-494566
EVASTAR ID: 1000049456

In der Reihe „Karlsruher Geochemische Hefte"
(ISSN 0943-8599) sind erschienen:

Band 1: U. Kramar (1993)
Methoden zur Interpretation von Daten der geochemischen Bachsedimentprospektion am Beispiel der Sierra de San Carlos/ Tamaulipas Mexiko

Band 2: Z. Berner (1993)
S-Isotopengeochemie in der KTB - Vorbohrung und Beziehungen zu den Spuren-elementmustern der Pyrite

Band 3: J.-D. Eckhardt (1993)
Geochemische Untersuchungen an jungen Sedimenten von der Galapagos-Mikroplatte: Hydrothermale und stratigraphisch signifikante Signale

Band 4: B. Bergfeldt (1994)
Lösungs- und Austauschprozesse in der ungesättigten Bodenwasserzone und Auswirkungen auf das Grundwasser

Band 5: M. Hodel (1994)
Untersuchungen zur Festlegung und Mobilisierung der Elemente As, Cd, Ni und Pb an ausgewählten Festphasen unter besonderer Berücksichtigung des Einflusses von Huminstoffen.

Band 6: T. Bergfeldt (1995)
Untersuchungen der Arsen- und Schwermetallmobilität in Bergbauhalden und kontaminierten Böden im Gebiet des Mittleren Schwarzwaldes

Band 7: M. Manz (1995)
Umweltbelastungen durch Arsen und Schwermetalle in Böden, Halden, Pflanzen und Schlacken ehemaliger Bergbaugebiete des Mittleren und Südlichen Schwarzwaldes.

Band 8: J. Ritter (1995)
Genese der Mineralisation Herrmanngang im Albtalgranit (SE-Schwarzwald) und Wechselwirkungen mit dem Nebengestein.

Band 9: J. Castro (1995)
Umweltauswirkungen des Bergbaus im semiariden Gebiet von Santa Maria de la Paz, Mexiko.

Band 10: T. Rüde (1996)
Beiträge zur Geochemie des Arsens.

Band 11: J. Schäfer (1998)
Einträge und Kontaminationspfade Kfz-emittierter Platin-Gruppen-Elemente (PGE) in verschiedenen Umweltkompartimenten.

Band 12: M. A. Leosson (1999)
A Contribution to the Sulphur Isotope Geochemistry of the Upper Continental Crust: The KTB Main Hole - A Case Study

Band 13: B. A. Kappes (2000)
Mobilisierbarkeit von Schwermetallen und Arsen durch saure Grubenabwässer in Bergbaualtlasten der Ag-Pb-Zn-Lagerstätte in Wiesloch

Band 14: H. Philipp (2000)
The behaviour of platinum-group elements in petrogenetic processes: A case study from the seaward-dipping reflector sequence (SDRS), Southeast Greenland volcanic rifted margin

Band 15: E. Walpersdorf (2000)
Nähr- und Spurenelementdynamik im Sediment/Wasser-Kontaktbereich nach einer Seekreideaufspülung – Pilotstudie Arendsee –

Band 16: R. H. Kölbl (2000)
Models of hydrothermal plumes by submarine diffuse venting in a coastal area: A case study for Milos, South Aegean Volcanic Arc, Greece

Band 17: U. Heiser (2000)
Calcium-rich Rhodochrosite layers in Sediments of the Gotland Deep, Baltic Sea, as Indicators for Seawater Inflow

In der Fortsetzungsreihe „Karlsruher Mineralogische und Geochemische Hefte“ (ISSN 1618-2677) sind bisher erschienen:

Band 18: S. Norra (2001)
Umweltgeochemische Signale urbaner Systeme am Beispiel von Böden, Pflanzen, und Stäuben in Karlsruhe

Band 19: M. von Wagner, Alexander Salichow, Doris Stüben (2001)
Geochemische Reinigung kleiner Fließgewässer mit Mangankiesen, einem Abfallsprodukt aus Wasserwerken (GReiFMan)

Band 20: U. Berg (2002)
Die Kalzitapplikation als Restaurierungsmaßnahme für eutrophe Seen – ihre Optimierung und Bewertung

Band 21: Ch. Menzel (2002)
Bestimmung und Verteilung aquatischer PGE-Spezies in urbanen Systemen

Band 22: P. Graf (2002)
Meta-Kaolinit als latent-hydraulische Komponente in Kalkmörtel

Band 23: D. Buqezi-Ahmeti (2003)
Die Fluidgehalte der Mantel-Xenolithe des Alkaligesteins-Komplexes der Persani-Berge, Ostkarpaten, Rumänien: Untersuchungen an Fluid-Einschlüssen

Band 24: B. Scheibner (2003)
Das geochemische Verhalten der Platingruppenelemente bei der Entstehung und Differenzierung der Magmen der Kerguelen-Flutbasaltprovinz (Indischer Ozean)

Ab Band 25 erscheinen die Karlsruher Mineralogischen und Geochemischen Hefte bei KIT Scientific Publishing online unter der Internetadresse http://www.ksp.kit.edu
Auf Wunsch sind bei KIT Scientific Publishing gedruckte Exemplare erhältlich („print on demand“).

Band 25: A. Stögbauer (2005)
Schwefelisotopenfraktionierung in abwasserbelasteten Sedimenten - Biogeochemische Umsetzungen und deren Auswirkung auf den Schwermetallhaushalt

Band 26: X. Xie (2005)
Assessment of an ultramicroelectrode array (UMEA) sensor for the determination of trace concentrations of heavy metals in water

Band 27: F. Friedrich (2005)
Spectroscopic investigations of delaminated and intercalated phyllosilicates

Band 28: L. Niemann (2005)
Die Reaktionskinetik des Gipsabbindens: Makroskopische Reaktionsraten und Mechanismen in molekularem Maßstab

Band 29: F. Wagner (2005)
Prozessverständnis einer Naturkatastrophe: eine geo- und hydrochemische Untersuchung der regionalen Arsen-Anreicherung im Grundwasser West-Bengalens (Indien)

Band 30: F. Pujol (2005)
Chemostratigraphy of some key European Frasnian-Famennian boundary sections (Germany, Poland, France)

Band 31: Y. Dikikh (2006)
Adsorption und Mobilisierung wasserlöslicher Kfz-emittierter Platingruppenelemente (Pt, Pd, Rh) an verschiedenen bodentypischen Mineralen

Band 32: F. I. Müller (2007)
Influence of cellulose ethers on the kinetics of early Portland cement hydration

Band 33: H. Haile Tolera (2007)
Suitability of local materials to purify Akaki Sub-Basin water

Band 34: M. Memon (2008)
Role of Fe-oxides for predicting phosphorus sorption in calcareous soils

Band 35: S. Bleeck-Schmidt (2008)
Geochemisch-mineralogische Hochwassersignale in Auensedimenten und deren Relevanz für die Rekonstruktion von Hochwasserereignissen

Band 36: A. Steudel (2009)
Selection strategy and modification of layer silicates for technical applications

Band 37: E. Eiche (2009)
Arsenic mobilization processes in the red river delta, Vietnam: Towards a better understanding of the patchy distribution of dissolved arsenic in alluvial deposits

Band 38: N. Schleicher (2012)
Chemical, Physical and Mineralogical Properties of Atmospheric Particulate Matter in the Megacity Beijing

Band 39: H. Neidhardt (2012)
Arsenic in groundwater of West Bengal

Band 40: X. Tang (2014)
Separating arsenic oxyanions from natural waters for oxygen isotope analysis

Band 41: A. K. Nothstein (2016)
Selenium Transfer between Kaolinite or Goethite Surfaces, Nutrient Solution and *Oryza Sativa*